THE LONG DRAG

A SHORT HISTORY OF BRITISH TARGET TOWING

DON EVANS

FLIGHT
RECORDER
PUBLICATIONS

A passion for accuracy

First published in Great Britain in 2004 by
Flight Recorder Publications Ltd
Ashtree House, Station Road, Ottringham,
East Yorkshire, HU12 0BJ
Tel: 01964 624223 Fax: 01964 624666
E-mail: beketley@dircon.co.uk
Website: www.flight-recorder.com
© 2004 Flight Recorder Publications Ltd

ISBN 0 9545605 4 X

Edited by Barry Ketley
Design by Flight Recorder Publications Ltd
Colour profiles by David Howley
Printed in England

Distribution & Marketing
in UK & Europe by
Midland Publishing
(a part of the Ian Allan Group)
4 Watling Drive, Sketchley Lane Industrial Estate,
Hinckley, Leics, LE10 3EY
Tel: 01455 233747 Fax: 01455 233737
E-mail: midlandbooks@compuserve.com

Distribution & Marketing
in USA by
Specialty Press
39966 Grand Avenue, North Branch, MN 55056,
USA
Tel: (001) 651 277 1400 Fax: (001) 651 277 1203
E-mail: davida@cartechbooks.com

ALSO AVAILABLE

KURT TANK'S PHOTO ALBUM 1940-1943
Compiled by
Roy Powell & Barry Ketley
ISBN 0 9545605 3 1

THE WARLORDS
US Eighth Air Force Fighter Colours of World War II
Volume 1
The 4th, 20th & 55th Fighter Groups
by
Barry & Ann Money
ISBN 0 9545605 1 5

A CIVILIAN AFFAIR
A Brief History of the Civilian Aircraft Company of Hedon
by
Eduard F. Winkler
ISBN 0 9545605 0 7

FORTHCOMING

CHECKERBOARD HUNTERS
The Croatian Volunteers of 15./JG 52 on the Eastern Front
by
Marko Jeras, Dragisa Brăsnović
& Zdenko Kinjerovac
ISBN 0 9545605 2 3

RISE & DEFEND
The USAAF at Manston
1950-1958
by
Duncan Curtis
ISBN 0 954605 5 8

ACKNOWLEDGMENTS
and an appeal
Photographs are from the author's collection, kindly supplemented by others from the collections of Cobham Plc via Colin Cruddas, Tim Mason, Gary Madgwick and Barry Ketley.

Compared to most other aviation activities, very few pictures seem to have been taken of target towing aircraft in service, particularly during World War II and earlier. The publisher would be pleased to hear from anyone with such pictures to loan, sell or exchange.

Caption to front cover: Canberra TT Mk 18 WK143 in service with Flight Refuelling Ltd dropping a development version of the Rushton target, which appears to be fitted with a flare pack, from the starboard Rushton winch.

Caption to title page: A rare inflight picture of the Miles Monitor NP407, the second production aircraft, taken on its first flight on 5 April 1944 in the hands of Flt Lt Tommy Rose. Its handsome lines and target tug stripes are readily apparent. In plan view the aircraft wore an astonishing similarity to the German Focke-Wulf Ta 154 night fighter. Both designs shared a similarity in being of mostly wooden construction, and both were fated never to be successful in their intended roles.

CONTENTS

INTRODUCTION
The long drag

The definition of a target according to most dictionaries is 'A butt or mark set up for the purpose of marksmanship or an aiming point'.

Ever since the day when primitive man made his first sling shot, spear, catapult or bow and arrow, he must have spent a considerable amount of time practicing against a tree or similar mark, and for every type of weapon that has been produced since that time, a target of some sort will have been used for the marksman to practice and to demonstrate his skills.

Throughout history there have been tournaments and competitions where marksmen have pitted there skills against each other, usually shooting at stationary targets, to decide who was the best marksman. However, the archer who could plant his arrow in the bullseye of a rush target ten times out of ten would have found it a totally different proposition to hit a running deer at the same distance. Obviously, he would have to aim at a point in front of the creature, so that by the time the arrow had covered the distance, the deer would have run far enough forward for the arrow to find its mark. The faster the deer was running, the further in front would be the aiming point. Knowing exactly how far in front the aiming point would be could only be determined by practice, and no amount of practice at a stationary target would improve his aim. It therefore follows, that whatever weapon a marksman may be using, if he is to become proficient, he must have a realistic target to practice on. It would be an utter waste of time for an anti-tank gunner to practice his shooting at a stationary target set at one hundred yards range, if, when firing in anger he would be expected to hit a vehicle a thousand yards away travelling at forty miles and hour. A third-rate clay pigeon marksman would probably knock down more pigeons than the Bisley champion.

With modern complex weapons, the provision of a realistic target is becoming more complex than the actual weapons. It is a comparatively simple task to provide a target for a rifle or anti-tank gun, but to provide something to cater for an aircraft, anti-aircraft gun or missile provides enormous problems. A highly sophisticated piece of equipment is required if it is to be of any real use in both the training and proving of the weapon. It is with this equipment and its development that this survey is concerned.

For many years, the aerial towed target was the principal method used, and the development of these targets and their towing aircraft were sorely neglected. Target Towing units were always regarded as the Cinderella of the Armed Forces, indeed, it was almost thought of as punishment to be posted to a Target Towing Squadron, by both groundcrew and aircrew alike. Invariably, the aircraft that were used were only engaged on Target Towing when they were obsolete and the Service had no further use for them.

3 Above: WJ680 is a Handley-Page built Canberra B Mk 2 seen on 31 October 1980 after it had been converted to a TT Mk 18. In service with 7 Squadron it still carries grey and green tactical camouflage and toned-down national markings on the uppersurfaces, but with the full black and yellow target towing colours on the underside. Note 7 Squadron's 'plough' seven-star emblem on the fin. There is a Rushton winch under each wing, each carrying two towed targets. There also appears to be some kind of guard or deflector under the tailplane root.

At the outbreak of war in 1939, much of the Target Towing was being carried out by Westland Wallace and Fairey IIIF aircraft which had first entered service in the late 'twenties. These were staggering around the sky at about 80 knots with a sleeve or banner target for the gunners to practice on. When the German bombers arrived, they were flying at twice the height and at speeds of over two hundred knots. Small wonder our gunners met with little success.

The position did not improve very much right through the war, and it was not until the early 'fifties that serious thought was given to a realistic target, although better target tugs in the way of the Defiant, Henley and Martinet appeared during the war. Since World War II, large sums of money have been spent on the development of new towed target systems and towing aircraft.

To appreciate some of the problems involved in designing a modem target system, one must think of a surface-to-air missile and its capabilities. It will probably be designed to destroy an enemy aircraft travelling at six or seven hundred miles an hour at an altitude of over forty thousand feet, and would be fitted with a heat seeking or radar device to home it on to it's target.

If a target is to be provided for such a weapon, then it is obvious that it must be capable of flying at similar speeds and heights. It must be capable of a heat source to simulate an aircraft jet exhaust and give off a radar reflection comparable to that of a full sized machine. Even more important, it is still a target, made to be destroyed and must therefore be cheap and consumable. A 'catch 22' situation.

One way of providing such a target is the use of Drone Targets. These are normal aircraft which have been rebuilt with radio control, being able to fly without a crew. Most of these are more expensive to produce than the original aeroplane, and with the accuracy of modern weapons, the cost of these could be prohibitive. There was an occasion during the 'fifties when the Royal Navy shot down seven Meteor U16s in a single day over Cardigan Bay with Sea Slug missiles. Questions were asked in the House...

Certainly, the Drone Target performs a valuable role in the development of the surface to air missile, but what of the ordinary bread and butter role of target practice? It is useless to produce a new weapon, prove that it works, then put it on the shelf until it is needed. Crews have to be trained and kept in training. This can only be achieved by constant target practice.

The answer lies in the use of the towed target. A cheap and expendable target towed through the air behind an aircraft. This has basically what has been used from the start, but modern requirements in performance present enormous difficulties. To start with, the towing aircraft must be capable of matching the speeds and altitude required, despite the weight and drag imposed by the addition of a towed target system.

It is generally accepted that the target should be six or seven miles (9.7 to 11.3 km) behind the towing aircraft when flown against homing missiles, as, should the homing source on the target fail after the missile has been released, then the missile would probably home on to the towing aircraft if a shorter tow length was used. The length of tow cable used to provide this distance will weigh several hundred pounds, and added to this is the weight of the target,consequently it will be seen that the load on the cable will be enormous, even before adding the drag which high speed flight will impose.

The target will have to be stowed on the aircraft and launched when required. A winch will be used to pay out the cable, and here we meet the next problem. If the cable is winched out at a thousand feet (305 m) per minute, which is a very high speed indeed, it would take almost an hour to winch out fifty thousand feet and a similar time to winch it back in again. Almost two hours of the aircraft flying time spent on winching could leave a very short time on task with a target.

The target has to be fitted with 'homing devices' to attract the missiles, and these would have to be activated by some form of signal from the towing aircraft which adds another complication. Systems currently coming into service are fitted with miss distance indication so that missiles can be offset to miss the target and the actual missile course recorded. All this, and yet the target has to be expendable.

Here then are some of the problems facing the designers. There are many more. In the following chapters we shall try to examine the history of the towed target, some of the problems involved and the attempts made to solve them.

4: A front view of a Rushton winch under the port wing of a Canberra TT Mk 18. The cylindrical containers probably house sleeve targets. Note the propeller mechanism which drives the winch in flight.

THE HISTORY OF AERIAL TARGETS
Flags, banners, sleeves and darts

How did it all start? Despite a considerable amount of time spent delving into records, very little reference to the subject has been found. There are many books available on almost every aspect of aviation but, apart from the occasional statement that this type of aircraft or another had been used for target towing, the details of how it was done and what equipment was used appears to be much too mundane to be recorded. Indeed, during my enquiries, I found that most manufacturers seemed to look on it as a disgrace if one of their aircraft had been used for target towing and did not want to admit it. For example, I wrote to the late Harald Penrose to get information about the Lysander Target Tower. In his reply he assured me that the Lysander had *never* been used for target towing. I could hardly argue with the great man, but *I* worked on them.

We know that the German Zeppelins of World War One had a system of target practice. The gunner threw a small parachute out of his gondola, and when it was streaming behind would do his best to shoot it to pieces. Sleeve targets were apparently used during the First World War for air-to-air gunnery, but these must have been quite small as a modern 4 x 20 sleeve target puts up a drag of around 300 lbs (136 kg) at 140 knots, and this would have been simply too much for the performance of the aircraft of the time.

During the 'thirties, when the first towed banners, appeared, I remember as a child seeing an elderly Avro 504 staggering along over Bournemouth beach dragging an advertising banner urging people to take Dr. Beecham's pills.

In later years I often wondered if some Service character saw it and thought what a marvellous target it would make? Or did some advertising genius see a target being hauled and say, "What a hell of a way to advertise"?.

During a conversation with the late Sir Alan Cobham, who was responsible for the advertising Avro over Bournemouth, he told me of one occasion when, with a banner hooked up behind the old Avro, it took him 15 minutes to cover the length of Bournemouth beach in a strong wind with an airspeed of 45 knots. He said that every flight was an adventure. The system used was quite primitive with a release hook attached to the aircraft which was operated by a length of cord which went up into the cockpit. If the aircraft ran into trouble, the pilot just heaved on the cord, thus releasing the rope and the banner.

At the start of a sortie, the aircraft was faced into wind with the rope laid out in a straight line behind, the rope being about 200 ft (61 m) long. A typical banner was made from canvas, about 3 ft (0.9 m) wide and about 20 ft (6 m) long. The canvas was attached to a metal spreader bar in the front with a pair of pram wheels attached to allow it to roll over the ground. Sometimes, an extra wooden spreader bar was attached at the rear to prevent the banner curling in flight. Four rigging lines were attached from the rope to the spreader bar, these being arranged asymmetrically to allow the banner to remain vertical in flight.

When all was ready, the aircraft would take off in the normal way, dragging the banner behind it. Very often, the banner was in a sorry state by the time it was airborne due to its rough passage across the grass. Frequently, the banner was so badly damaged during take off, the pilot would have to drop it, land, hook up and start all over again. Sir Alan told of one occasion when they made seven attempts to get it airborne, and when success was eventually achieved, the rope broke on the first run and the banner dropped into the sea off Swanage.

Certainly, the RAF were using a type of banner target for air-to-air firing behind an Avro 504K in 1929-30, but little banner flying was done before that time. Even so, banners were the first type of target to be generally used by the RAF and have had a long life, still in use today. They are only practical with comparatively short tow lengths (800 ft/244 m) and are used mainly for air-to-air firing, although the Mk 2 Banner Target could be trimmed to fly flat for ground-to-air low-altitude gunnery. During 1933, Flt Lt C.W. McKinley-Thompson from the Experimental Establishment at Martlesham Heath carried out a large number of trials involving a large banner target. The results led directly to the requirement for eight-gun fighters for the RAF.

When most people think of the aerial towed target they immediately think of the sleeve or drogue. This consists of a cylinder of fabric which resembles the wind sock used for showing wind direction on airfields. It is difficult to say when the first sleeve target was used, but I have seen a photograph showing a Blackburn Baffin aircraft towing a sleeve over Martlesham Heath, allegedly in 1931-1932. Like the banner, the sleeve is now only in limited use, mainly for elementary gunnery.

The sleeve had a big advantage in that it could be folded up into a small package and air launched from the towing aircraft, thus avoiding the circus-like dragging-off of the banner. In the early stages, the sleeve target was pushed out of the aircraft through a hole in the floor and the towing rope was packed into a dispenser. As the rolled target fell from the aircraft, it unrolled from its package and deployed to its full shape, the drag of the target serving to pull the rope out of the dispenser. The length of tow was still limited using this method and the doubtful accuracy of the gunnery in those days made target towing a rather hairy occupation with the target so close to the aircraft.

The use of cord or rope for target towing had another serious disadvantage, this being the drag. It is not generally appreciated how much drag can be caused by a rope in the air. Even a modern tow rope of $^{7}/_{16}$ in (11 mm) diameter with a breaking strain of 3,500 lbs (1,587 kg) and a length of 8,000 ft (2,438 m) will produce a drag equal to the breaking strain of the rope at 400 knots. The logical answer was to use a flexible steel cable as a tow line. Wire, however, could not be packed into a dispenser in the same way as a rope. It tends to kink, and a kink in a cable means a certain break.

The answer to this was the use of a winch. Several thousand feet of cable could be wound on to a drum and this could be paid out to the length required. The first of this type of system consisted of a drum with the cable wound on with a brake to stop the drum when required. At the end of the sortie, the brake was released allowing the cable to run off the drum and fall away. Later winches were fitted with an air driven turbine, allowing the cable to be winched in again after the sortie. Many types of winches have been used over the years and although most have been driven by a propeller or air turbine, there have been examples of winches driven hydraulically or electrically.

Towards the end of World War II, several attempts were made to produce a more realistic type of target — the era of the winged target had arrived. Several of these were produced on both sides of the Atlan-

6 Below: *High in the blue in March 1958, Meteor TT Mk 20 WD767 streams a sleeve target. The small bullet-shaped device on the cable is a SAAB miss distance indicator. These apparently suffered from intermittent electrical connections to the recording equipment in the aircraft.*

tic, being basically towed gliders. None of these, however, were an outstanding success, all being found expensive to produce, difficult to maintain and temperamental in flight. The slightest damage would cause them to go unstable, and when this happened the cable would have to be cut with the subsequent loss of the target. Despite their problems, some winged targets were produced in large numbers, particularly in the USA, and some of these were in use as late as 1950.

Most of the winged targets used by the RAF were used in conjunction with the winch-equipped aircraft, being towed off the runway on a short tow length and reeled out to full towing distance once airborne. The targets were trimmed to fly slightly lower than the aircraft, and, at the end of a sortie, the aircraft would fly in as if for a landing, and as soon as the nose wheel of the target made contact with the ground, a cable cutter would operate, separating the target from the aircraft, which would then do another circuit before landing. As the target nose wheel touched the ground, it would turn about 10° which turned the target off the runway making room for the tow aircraft when it landed.

After the winged targets came the 'darts'. The dart target was introduced mainly to allow higher towing speeds as the small frontal area presented a very low drag. The darts were relatively cheap to produce, had none of the temperament of the winged target and were capable of absorbing a considerable amount of punishment without going unstable.

There were several types in use with the British Services, all consisting basically of either three or four plywood fins fitted around a metal spine with a towing arm or stirrup fitted at the centre of gravity.

The latest of these was the Mk 3 which was designed to be air launched from a Meteor TT20 aircraft using the 'G' type winch and a 15 cwt (762 kg) steel tow cable. The Mk 3 had three fins set around

the central spine at 120°, and when the target was stowed on the aircraft, the lower fin folded up laying close to one of the upper fins. As the target was launched from the aircraft, the lower fin dropped and locked into position. The target was attached to the towing cable by a 20 ft (6 m) length of nylon cord. Immediately after launching, the target would fly on the nylon cord until the winch was set to pay out to the required tow length. After the sortie, the target would be winched in to the end of the cable, then released. A small recovery parachute was then released from the tail of the target which allowed the target to be recovered for future use.

The Mk 1 and 2 dart targets were more basic and were thus expendable. The method of getting these targets into the air was by using the highly spectacular air snatch technique. Aircraft such as the Sea Hawk, Sea Vixen, Scimitar or Canberra were fitted with a long hooked pole which trailed from the underside of the aircraft. In the case of the naval aircraft, the pole was attached to the arrester hook. The aircraft flew in at a height of 15 ft (4.6 m) and engaged its hook into a nylon loop suspended between a pair of poles. Attached to the loop was a 4,000 ft (1,220 m) or 6,000 ft (1,829 m) length of nylon tow cable with the target attached to the other end. As soon as the hook engaged the rope, the aircraft would go into a steep climb dragging the rope and target almost vertically into the air. At the end of the sortie, the aircraft would drop the hook, rope and target at a pre-arranged dropping zone. The hook and rope could be used again but the target usually penetrated 6 ft (1.8 m) into the ground.

The snatch system was used for several years but was eventually discontinued, mainly due to the large snatch and dropping area required and the lengthy process of setting up prior to the snatch and the necessity of a ground controller in touch with the aircraft during the snatch and drop. The tow length was

7: *Seen at Hamble, this is an example of the Blackburn and General Aircraft winged target. It differs in several details on the rear fuselage and spoiler from later models. (See photo 31)*

restricted to a maximum of 6,000 ft (1,829 m) and the excercise called for good weather conditions

After the war, with the introduction of jet aircraft as tugs with their limited endurance, time in the air became of paramount importance and the air launch became very desirable. This new generation of towing aircraft presented problems as far as the air launched banner was concerned. As they were unable to reduce their flying speed below 130-150 knots, the existing tow ropes were unable to stand the strain of the sudden jerk when the banner was launched at this speed. Fortunately, the introduction of nylon ropes with their great strength and elasticity provided fresh hopes for the air launched banner.

After several years of experimenting with various types of aircraft and systems, a successful pack was produced by the RFD Company of Godalming and this was subsequently fitted to a Sea Hawk aircraft with promising results. A development of this system was designed around the new Mk 3 Radar Responsive Banner and this was fitted to a Hunter F. Mk 6 aircraft which was extremely successful. The Hunter carried a container under each wing so that in the event of a target being 'shot up' it could be jettisoned and a second one launched.

The RFD Banner Target Container is manufactured from light alloy and is oval in cross section being fitted with an upper and lower compartment. The

upper compartment carries the webbing link and the 800 ft (244 m) length of nylon rope mounted on a tray or dispenser. The lower compartment carries the banner target and a spring ejector system for launching. The target is held against the spring tension by an electro-mechanical release slip When the release slip is operated, the spring throws the target rearwards into the airstream drawing after it the webbing link and tow rope. At the forward end of the tow rope, the steel cable is attached to an electro-mechanical slip which is fitted centrally under the fuselage. The

target is towed from this release hook, and at the end of the sortie the operation of this release frees the target and rope from the aircraft.

RFD's airborne release banner system was used on Hunter F.Mk 6 and Canberra TT.Mk lS aircraft. Although almost any type of aircraft can be used for banner towing, it has been the practice to use fighter aircraft for this purpose. Meteor, Hunter, Sea Hawk, Sea Vixen and Javelin aircraft have all been used from time to time. Very little modification is required to prepare an aircraft for towing, the main addition being the fitting of an electro-mechanical release hook. There is considerably more work involved when fitting an airborne release pod, and this includes the fitting of a twin release hook, one for each banner.

There are several types of banner target still in use today. They are all basically similar in appearance and some of the more common types are as follows:

Target Banner Type A-6A (US Air Force)

The American A-6 target is approximately 30 ft (9 m) long and 6 ft (1.8 m) in width. It is made from a light mesh of steel wire covered with cotton thread. The wire provides strength to the target and facilitates radar tracking. Towing of the target is by a bridle consisting of four webbing straps secured around the tubular steel spreader bar. It can be towed in either a vertical or horizontal position by changing the arrangements of a counterweight. This weight is pinned to the lower end of the spreader bar for vertical towing, and for horizontal towing it can be detached and fitted to an arm in the centre and bottom of the spreader bar.

Their target is launched by the snatch or drag technique and is not cleared for speeds in excess of 180 knots as severe whipping of the tail will take place and this will cause the mesh to disintegrate.

The 30 ft Banner Target, Radar Responsive, Mk 2

The Mk 2 Radar Responsive Banner Target is designed for towing speeds of up to 200 knots. It consists of a fabric net and an 8 ft (2.44 m) towing bridle which is attached to a tubular steel spreader bar, weighted at one end so that it adopts a vertical position in flight. The fabric is woven from a polythene filament and a metalising element forming a net of six strands to the inch.

The banner is 30 ft (9.1 m) long and 6 ft (1.8 m) in width. The towing bridle consists of seven 400 Ib (181 kg) nylon shroud lines and one length of 700 Ib (318 kg) nylon cord which is attached to the bottom of the spreader bar. The spreader bar has a cast iron weight of 171lbs (7.7 kg) attached to the lower end and a Sinch steel disc at the other. It was found during firing practice that the steel spreader bar was vulnerable when hit, consequently modification TT7 was introduced which provided a 2 inch (50 mm) diameter light alloy tube as a secondary spreader bar.

The 30 ft Banner Target, Mk 4

The Mk 4 Banner Net Target is basically the same as the Mk 2 previously described. The only difference is in the material used. The fabric is woven from a nylon filament without the metalising element. Being in itself non-radar reflective, this target was normally used with an aiming reflector.

The 30 ft Banner Target, Radar Responsive, Mk 3

The Mk 3 Banner Target is constructed from a rectangular sheet of open weave nylon measuring 30 ft (9.1 m) by 6 ft (1.8 m). The fabric is attached to a tubular glass fibre former which houses an internal light alloy radar reflector and a stabilising weight at the lower end. Used principally as an air launched target, and when carried in a container, the lines of the towing bridle are stowed in looped pockets of 2 inch (50 mm) webbing which are stitched to the leading edge of the banner immediately in front of the spreader bar.

Should the Mk 3 be required for the snatch or drag technique, removable side wheels are available. The total weight of the target is 40 lbs (18 kg) with a further 6 Ibs (2.72 kg) if the side wheels are added

The Mk 3 can be flown up to 250 knots and can be operated up to 30,000 ft (9,144 m).

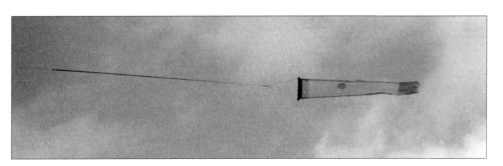

10: A rare picture of a banner target in actual use; here on 26 May 1981 being towed by a Canberra TT Mk 18 to give Lightning jockeys from Binbrook an opportunity to use their 30 mm Aden cannons.

SLEEVE TARGETS
Dragging a drogue

When most people hear aerial targets being mentioned, their minds usually turn to the sleeve or drogue. This is not surprising as the sleeve was one of the first type of target to be used, and later versions are still in service today. Over the years the sleeve has been the most common target to be used throughout the world.

Sleeve targets are nothing more that cylinders of fabric which can be folded up into small packages until required for use. When they are launched into the airstream they fill up with air which forms the target into its full shape and size

Although some sleeve targets are purely cylindrical, these being mainly American in origin, the majority have taken the form of a truncated cone. The earlier versions had the larger end in the front, but the later low-drag versions had the smaller end in the front with the larger end partially filled in. Whatever the shape, all of the sleeves have had a semi-rigid ring stitched in to the front end to hold the mouth of the sleeve open. A towing bridle consisting of a number of rigging lines is stitched to this ring, with the lines being tied at equal distance around its circumference.

It is difficult to say when the sleeve target first came into use but certainly they were being used widely by the RAF in the early thirties. A photograph in a Service magazine around this period shows a sleeve target, referred to as a Mk 1 Air Target Drogue being towed by a Westland Wallace aircraft. This target measured 2 ft (60 mm) in diameter at the front end and 1 ft (30 mm) at the rear, the overall length being 12 ft (3.66 m). It was towed on 1,000 ft (305 m) of cord and was used for both air-to-air and surface-to-air firing practice.

By 1939, the low drag target had come into being. The low drag was achieved by reversing the taper of the cone and partially blanking off the larger end which was at the rear. With this design, the sleeve inflated like a balloon forming a comparatively clean shape to present to the air stream. This basic design has been followed on all subsequent sleeve targets and modern sleeves differ only in the size, proportion and the material used.

The great advantage of the sleeve target, apart from its cheapness is its ability to be rolled into a small package which can be ejected from the aircraft, breaking out and deploying when it reaches the end of the towline or halyard. Even before the advent of winches it was possible to carry several sleeves in the aircraft so that in the event of a target being damaged, another could be pushed out through a hole in the aircraft floor. The replacement could be slid down the towline and knocked off the damaged target when it reached the end.

With the introduction of target winches, the sleeve target really came into its own, being air launched and paid out on several thousand feet of flexible steel cable thus allowing for higher towing speeds and longer tow lengths.

As with banner targets, the introduction of radar in attack systems called for the target to be radar responsive. This was provided in sleeve targets by the addition of a diaphragm of metallised mesh stitched into the sleeve at the rear end. Some American sleeves used longtitudinal and transverse panels of wire bearing cloth which were stitched to the inside of the sleeve throughout its length.

Sleeve targets in use in this country are manufactured from a rubberised nylon fabric known as Hyperlon and are available in three standard sizes, 2 ft x 10 ft (0.6 m x 3 m), 3 ft x 15 ft (0.9 m x 4.5 m) and 4 ft x 20 ft (1.2 m x 6 m). The first figure relates to the diameter at its widest point and the second figure to its length. The 4 ft x 20 ft target is seldom used these days, its main purpose is for higher altitude visual firing. The 2 ft x 10 ft sleeve is also not in general use and is usually used to lay a new cable on to the winch to ensure that the cable is in phase with the laying on gear.

When the 3 ft x 15 ft sleeve is prepared for use it is contained in a package measuring 2 ft 6 in (760 mm) in length and 6 inches (150 mm) in diameter. When the target is launched from the aircraft at a speed of around 140 knots, the package breaks open and the target fills with air taking up its full shape. This unfolding, or deployment to use the correct term, takes place in approximately one fifth of a second and is accompanied by a loud report like an explosion. This sudden deployment subjects the cable to an enormous shock load which could cause the breakage of the cable.

The figure of 140 knots is about the lowest speed at which a modern aircraft can operate and, obviously, higher speeds would impose even greater loads.

To overcome this problem, a nylon cord or halyard is fitted between the target bridle and the tow cable. The halyard, which is about 100 ft (30 m) long acts as a shock absorber due to the elastic qualities of the nylon. When preparing a sleeve target for use, the halyard is folded into a fabric stowage pack or envelope. The loaded envelope is wrapped around the folded target forming the outer layer of the delivery package.

Despite the flimsy appearance of the sleeve it is surprising that it can be towed at higher speeds than the banner although it is less robust when it comes to absorbing damage by gunfire. Even quite small holes in a sleeve will allow the air to spill out and cause it to cavort in an alarming manner. Fortunately, sleeves can be replaced easily in flight so that their vulnerabiliy does not present a major problem.

The sleeve target is the only type of target which can be successfully exchanged in flight and over the years, many interesting and various methods have been tried, some of which will be described later.

There is a tendency for sleeves to rotate in flight and this could have the effect of twisting up the tow cable, which could of course cause a breakage. To overcome this problem a ball bearing swivel is fitted between the tow cable and halyard which allows the sleeve to rotate freely.

Sleeve targets have been with us for many years, and even in this age of high speed targets and guided weapons, the sleeve is still used for gunnery training and probably will be for many years to come.

12 Below: A diagram from the relevant Air Publication (AP 1492) showing the correct method of folding a sleeve target, and a sketch showing its shape when deployed.

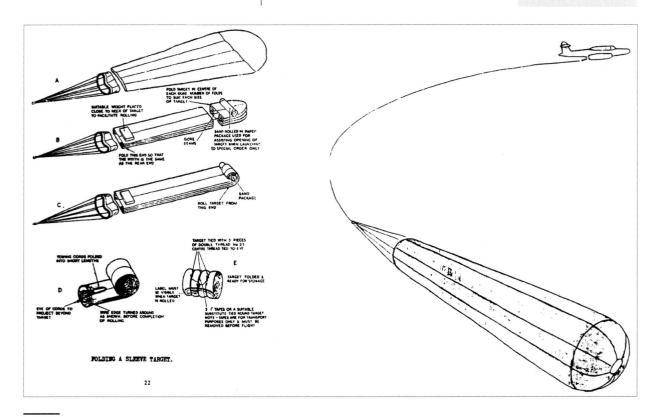

FOLDING A SLEEVE TARGET.

22

TARGET EXCHANGING
Sleeve swapping

This most interesting subject deserves a chapter of its own. For a period of over fifty years, to the target towing crews, this was what target towing was all about. Mainly concerned with sleeve targets, the purpose of the exchange was to replace a target which had been damaged or destroyed by gunfire with a new target without the aircraft having to return to base. With some installations, targets could be exchanged several times during a single sortie.

With the wartime aircraft such as the Henley, Defiant and Martinet using the 'B' Type winch, the operator sat in the rear cockpit with the winch installed transversely across the fuselage above his knees. The sleeve targets were folded as described in the previous chapter with the launching halyard attached and were neatly stowed around the cockpit being held in place by elastic 'bungee' cords.

On the initial launch, a target would be selected from its stowage and the end of the halyard was tied to the eye of the steel cable which was wound onto the winch. The halyard also had an eye spliced in the end and the method of tying was to pass three loops of 12 lb (5 kg) cord through the two eyes, finishing with a reef knot. (See sketch above).

When the target was attached, on the order from the pilot, the package was smartly thrown out of the aircraft through the launching hatch in the cockpit floor. As soon as the sleeve was deployed and was streaming from the aircraft on its halyard, the winch was set in motion and the cable paid out to the required towing length.

Attaching the halyard to the cable

When it became necessary to exchange the target, the winch was set to haul in and when the knot came up into the aircraft the winch was stopped. The knot was cut through with a sharp knife allowing the damaged target to fall away from the aircraft. Another sleeve was then selected, tied to the cable, launched and paid out. To assist the operator to know when the knot was about to appear in the cockpit, the last 100 ft (30 m) of cable was painted red and the final 20 ft (6 m) painted yellow. If, in the event of the painted portion of the cable being shot away, the operator had no idea of how much cable was left remaining.

One of the first automatic target exchangers was of American origin. This had the additional advantage of allowing the target to be exchanged without the necessity of winching in the cable first, which also saved a considerable amount of time. The exchanger unit which fitted onto the end of the cable was of metal construction and tubular in shape, the tube containing a latch and a trigger. The target halyard was attached to a ring which slid along the cable and over the tube. The ring passed over the trigger and was prevented from sliding off the end by the latch. Before the release unit was attached to the cable, the towing rings of the spare targets were threaded over the cable which passed through the rings when paying out.

When an exchange was required, the new target was launched through the target hatch in the normal way. The target and halyard dragged the ring over the cable until it reached the release unit at the end, and as it slid over the tube, the ring depressed the trigger which released the latch allowing the damaged target to fall away. As soon as the second ring had passed over the trigger, the latch was free to return in time to prevent the ring from sliding off the end of the exchanger.

This type of release and similar releases of British design worked extremely well but were limited to aircraft which could reduce their speed to 80 knots or less, as, at higher speeds, the sudden jerk as the target reached the end of the cable was sufficient to cause the cable to snap.

The British Type 'E' and 'F' winches which were fitted in Martinet and similar aircraft were electrically driven, but the electric motor lacked the power to reel the cable back in with the target attached. They were fitted with more than one drum of cable, and when the target needed exchanging, the complete target and cable was slipped from the drum and the next drum of cable was paid out with a fresh target. This system was very wasteful as the cable was often lost completely, and even if it was recovered, it was usually in such a tangle it was useless for further work

An interesting device was introduced to improve this situation. This was known as the Pembray Release Gear. This unit, which only weighed two and a half pounds was attached to the end of the cable of an 'E' or 'F1' winch. It consisted of a spring loaded release catch contained in a tube which held the catch in the closed position. To detach a target in tow, a small metal spool known as a messenger was released from the aircraft. This messenger slid down the cable striking the leading edge of the release gear on reaching the end. The resulting impact drove the tube rearwards against the spring causing the release catch to rotate and free the target.

Once the target had been released, the electric motor was capable of winching in the bare cable so that a fresh target could be attached. Using this system, only one drum was usually used and the proceedure repeated to use all of the targets carried in the aircraft, obviating the necessity of losing the cable every time a target was used giving a greater saving in cable and field work.

One of the most successful of all exchanger systems was the Type 'B' target exchanger used in conjunction with the remotely controlled 'G' winch. This combination was used on a variety of aircraft including Tempest, Beaufighter, Firefly, Mosquito and Meteor TT Mk 20. The last named was the last of the specialised target towers as later aircraft were being fitted with a long tow capability with sophisticated targets where sleeve towing was a secondary consideration.

The Meteor TT Mk 20 carried the 'G' type winch on top of a pylon mounted on top of the port mainplane, the winch being controlled remotely from the rear cockpit. The cable was routed out of the rear of the winch and carried over guide tubes and pulleys to the underside of the aircraft. It was then passed through a mechanically operated cable cutter and a spring loaded telescopic buffer strut. The sleeve targets were carried in four cannisters inside of the fuselage, two on either side and just behind the buffer strut. Each cannister had a door which was operated remotely from the operators cockpit. Each sleeve was packed in the normal way but had a jockey ring attached to the end of the halyard which was threaded over the end of the buffer strut and held in position by a spring clip. The four sleeves were then stowed in the cannisters and the doors closed. The target exchanger was then fitted to the end of the towing cable. (See diagram on page 17).

At the start of a sortie, the winch cable was pulled in tight holding the release unit against the buffer strut, thus compressing the spring. When the first target was required, the operator selected No. 1 target release which operated a solenoid catch allowing the No. 1 cannister door to open. The sleeve was then ejected into the airstream by 'bungee' elastic cords. As the sleeve came to the end of the halyard, it deployed and the resulting snatch pulled the jockey ring off the end of the buffer strut and onto the end of the target release unit. With the exchanger unit spring being compressed, the first set of pawls arrested the jockey ring preventing any further rearward movement. The winch was then set to pay out the cable. As the tension was relaxed the exchanger unit spring expanded allowing the unit to open up and release the first set of pawls. This action allowed the jockey ring to slide rearwards to be held by the rear set of pawls which opened as the first pawls closed. The target was then winched out to its operating tow length.

When the target was damaged necessitating an exchange, the winch was set to haul in and, as the end of the cable came up to the aircraft, the exchanger unit was pulled up against the buffer strut, compressing the spring and thus releasing the rear

14: *A sketch showing the working components of the American Target Release Mk 3. This was useable only by aircraft which could reduce their speed to 80 knots or less, as in the event of a target being changed at higher speeds the cable was liable to break.*

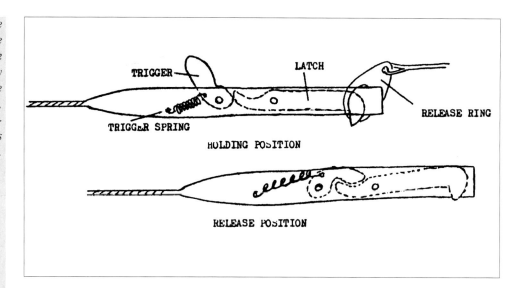

15: *Another target exchanger system was the British Type 'B', used in conjunction with the Type 'G' winch. This was fitted to aircraft as diverse as the Tempest, Beaufighter, Firefly, Mosquito and Meteor.*

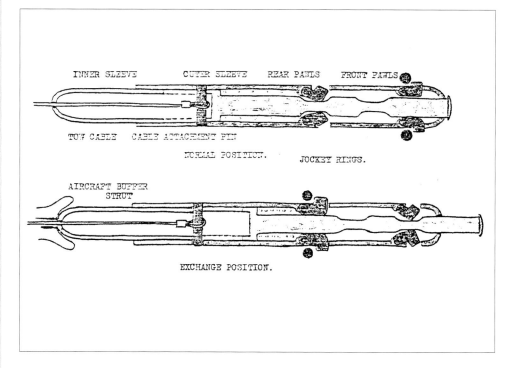

16: *This diagram shows the main workings of the A&AEE developed Rapid Target Exchanger. It worked extremely well, but was not used extensively on account of the introduction of the Delmar winch and Rushton target.*

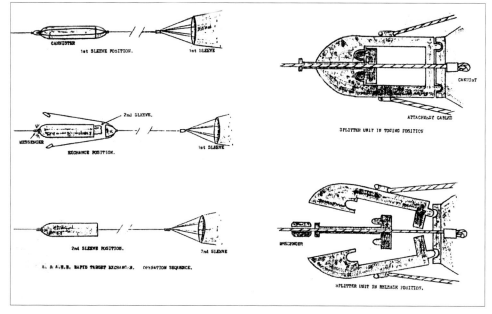

pawls. This allowed the sleeve to pull the jockey ring off the exchanger unit and fall away. The number 2 target was then selected and the whole operation repeated.

Great care had to be taken when preparing an aircraft for a sortie to ensure that the jockey rings for each target were threaded on to the buffer strut in the correct order, and naturally, the operator had to ensure that he selected the targets in strict rotation. If, by chance, the trap door of a particular target failed to release, the sortie would have to be cancelled as it would not be possible to select the next target instead. If this was done, the jockey ring would be held by the jockey ring of the previous target which had failed to release

The buffer strut in the system also fulfilled another purpose and provided an automatic control of the winch. When the winch was set to haul in, it was usually set to wind in the cable at 1,000 ft per minute (305 m per minute). The winching speed was set by varying the pitch of the windmill blades. When all but 200 ft (60 m) of cable was left out, a switch on the winch feathered the windmill and applied the brake. When this stage was reached, the operator selected the 'in override switch' which turned the windmill blades a few degrees of pitch and released the winch brake. This action allowed the winch to rotate slowly, hauling in the last 200 ft (60 m) of cable at 200 ft per minute until the exchanger unit reached the buffer strut thus compressing the spring. A micro switch was fitted to the buffer strut which applied the brake when the buffer strut was compressed.

The system described above was a great advance as the operator never had to handle the winch or the target, everything else being done by remote control. This was very much easier for the operator and would almost seem to be the ultimate in sleeve target operation. This was not to be, as another problem arose which would be a very difficult one to answer.

The Meteor TT Mk 20 was by then, the standard target towing aircraft in service and even with long range fuel tanks fitted it had an operating duration of only one hour and forty minutes. The length of time required to launch and pay out a target to the maximum 6,000 ft (1828 m), haul in to exchange, then pay out the second target meant that there was never enough time to use a third sleeve and a fourth sleeve was an impossibility. In fact Messrs. Airwork of Hurn, who were operating TT Mk 20s under contract, were even flying to the firing range in Lyme Bay on one engine to conserve fuel!

Something was needed to speed up the method of exchanging. It will be recalled that a much quicker system was being used on the Martinet some years earlier, whereby a second sleeve was sent down the cable which knocked off the first sleeve when it reached the end. At first, it was thought that a similar system could be used. It was simple in the case of the Martinet which could be flown at speeds as low as 80 knots, but alas, the Meteor could not be flown at much less than 140 knots, and the target sliding down the cable at this speed built up such a velocity that the cable snapped at every attempt.

The ML Aviation Company of White Waltham were old hands at the target towing game and were charged with the task of producing a rapid target exchanger. During the months that followed, many ingenious devices were produced, some of which worked on occasions but all of which were too inconsistent for Service use.

Finally, the Towed Target Development Section at Boscombe Down came up with an idea which proved to be highly successful and permitted sleeve targets to be exchanged in the incredible time of 20 seconds. In the following years no failure was ever recorded.

Up to this time all the thinking had been on the lines of sending the exchange target down the tow line. The Boscombe Down system broke away from this idea and worked on the principle of having two targets at the end of the cable right from the start with the first sleeve deployed on a 100 ft (30.5 m) halyard and the second one folded until required.

A streamlined fibreglass container was fitted to the end of the cable, 2 feet (60 mm) in length and 9 inches (230 mm) in diameter. The container was fitted with a detachable rear cone. The second, or exchange, target was packed with its halyard into the container with the halyard attached inside at the forward end.. The halyard of the first sleeve was attached to the detachable rear cone of the container

In flight, the container was stowed in a bracket beneath the aircraft with the towing cable attached to the nose cone which was pulled up into the buffer strut. The first sleeve was contained in the No l sleeve cannister. When launching, the first sleeve was released in the normal way and after deployment it flew from its halyard from the rear of the container.

The winch was then set to pay out to its operating tow length. When an exchange was required, a messenger in the form of a four ounce (113 gm) spool was released to travel down the tow cable. This messenger, although not heavy enough to snap the cable, had enough energy when it struck the front end of the container to release the mechanism. Operation of the release allowed the rear cone of the cannister to fall away with the first target, thus releasing the second target which deployed after leaving the cannister.

The twenty seconds exchange time was almost entirely the length of time taken by the messenger to slide down the tow cable, the actual time taken to exchange taking only one eighth of a second. The exchange was something I could never really get used to, even after the many trials carried out to prove the system. The actual exchange was too quick to be observed with the naked eye and one always thought one was still watching the first sleeve on the towline until it was seen falling away some distance behind.

Despite the success of the Boscombe Down Rapid Target Exchanger it did not see very much service, although several sets were made by the Royal Naval Dockyard at Perth. The reason was the introduction of the new Delmar Winch and sophisticated Rushton Target. This was the end of an era, and for me, the sport had gone out of target towing.

17: Miles Martinet TT Mk II RH121 was the last of a batch of nine for the RAF and RN. The projecting arm of the Type 'B' winch is clearly evident, although it is missing the windmill blades. Projecting from the lower fuselage is the support frame and pulley around which the cable to the sleeve target ran. A small 'D' shaped guard is just visible behind the tailwheel, fitted to prevent the cable fouling the rudder.

18: A diagrammatic view of the Type 'G' winch installation on the wing root of a Meteor TT Mk 20. The Delmar winch occupied the same position, but was considerably lighter.

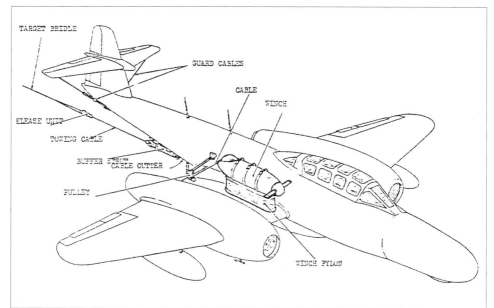

19: This view of Meteor TT Mk 20 WD706 in November 1962 shows the positioning of the Delmar DX4 winch on the wing root pylon and the various guard wires around the tail surfaces. Note the Argosy in the background. © Crown Copyright.

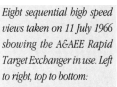

Eight sequential high speed views taken on 11 July 1966 showing the A&AEE Rapid Target Exchanger in use. Left to right, top to bottom:
20: The first sleeve in flight;
21: The striker releases the first target which is just breaking away.
22-24: The first target falls away.
25: The second target deploying from its canister.
26: Second target almost deployed.
27: Second target fully deployed. This whole sequence took place in less than ten seconds. What happened to all the used targets dropped around the countryside?

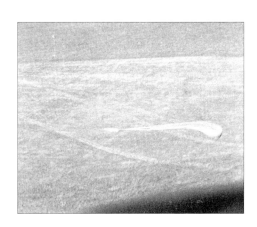

WINGED TARGETS
Less than perfect

Although winged targets are rarely, if ever, used these days, there is no doubt that they provided the most interesting of all towed targets. These machines were little more than gliders being towed from the tug aircraft, sometimes with a fixed tow rope of around 1,000 ft (305 m), or more often, by longer lengths of up to 6,000 ft (1,830 m) of cable when using a winch.

The reason for the introduction of winged targets was the requirement for a high speed target of realistic appearance and certainly, when flying as intended, they certainly provided this.

Several different types were produced on both sides of the Atlantic and they provided a useful service from around 1945 to 1955, although odd examples were probably around before and after these dates. There were a number of disadvantages shared by all types of winged target. They were generally temperamental to service and fly, they were unable to sustain damage and remain stable and they also required a large ground support effort to. prepare them for flight.

In an operation manual for the American type Aero 27A it stated, *"This target is a towed aerodynamic body with no intelligence aboard. It must fly by itself and no corrective action can be taken to aid its flight once it is airborne. Because of this, extreme care must be taken in assembly, repair and maintenance if there is to be any hope of* *satisfactory operation".* This quote sums up the temperament of all winged targets.

The Aero 27A was a high speed low drag metal target of peculiar design which could be operated at speeds of up to 450 knots. Designed and built by East Coast Aeronautics Inc., it was used in large numbers by the United States Air Force for several years. It wasa operated using the drag technique for take off with a 1,000 ft (305 m) nylon tow line, and when landing, the target was released automatically from the tow aircraft on contact with the ground, and simultaneously, an 8 ft (2.4 m) parachute was released from the upper aft fuselage to shorten the grond run and avoid damage to the target.

Most of the winged targets used a similar release system, being operated by the undercarriage, usually the nose wheel which slipped the tow line when the weight of the target compressed the wheel on landing. The targets were usually trimmed to fly lower than the aircraft, so that when landing,, the target touched the ground first and disconnected, allowing the tow aircraft to make another circuit before landing. The nose wheel of the target was so designed, that when the weight of the target bore down on the nose wheel, it automatically turned a few degrees thereby making the target roll off to the side of the runway giving a clear path for the tow aircraft to land after its extra circuit.

The most successful of the British winged targets

was the 'twenty-five footer'. Like its American counterpart, it was of all metal construction but was more orthodox in appearance and slightly larger.

The 25ft (7.6 m) winged target had a light alloy fuselage of monocoque construction and wings of alloy, fabric covered. It had a wingspan of 25 feet, a length of 26 ft (7.9 m) and an all-up weight of 106 lbs (48 kg). The target was usually used with a winch fitted aircraft being dragged off the ground on a 200 ft (61 m) halyard which was attached to the target towing bridle. Once airborne, the winch paid out the cable to its operating tow length and winched in again prior to landing.

There were several other types of winged targets produced in this country but very few were actually used by the Armed Forces. Worthy of note are the Blackburn and General Aircraft tailless target of which records are scanty but was certainly tested by the A&AEE at Boscombe Down, and the RFD Twin Boom target. This machine had a wingspan of 26 ft (7.9 m), a weight of 50 lbs (23 kg) and was constructed of resin bonded plywood. Due to the materials used in its construction it had to be fitted with a separate radar reflector and was also fitted with a braking parachute. Unfortunately, its maximum towing speed was only 240 knots which would have limited its use by the Services.

Although winged targets were normally taken off using the drag technique, a system was tried out launching the target from the upper mainplane of a Swordfish or Shark aircraft during flight, both machines being fitted with a 'B' Type winch. Although this system appeared to work it was not used generally in service, either due to the shortage of biplane tugs or to the hazardous nature of the launch.

The majority of winged targets arrived at the squadrons in packing cases and had to be assembled and rigged by experts. They had to be treated as real aircraft and special tools were needed, as were jigs and rigging boards. When in flight, they were very prone to damage which invariably caused the target to go unstable and in these cases they would have to be jettisoned.

Despite the fact that the winged targets were popular with the gunners, being the most realistic target with the exception of the drone, they fell out of favour due to their temperament and operating costs. Notwithstanding all these things, winged targets, in their day, provided one of the most interesting chapters in the story of the towed target.

29: A real rarity! A Westland Walrus (a hideous derivative of the DH 9A) carrying an air-launched target on the upper wing for trials at Gosport in May 1925. The target bears a passing resemblance to a German Rohrbach flying boat...

30: This twin-boom affair is an RFD winged target. It was not adopted for service use as it could not be towed at a high enough speed.

31: *Another view of a Blackburn & General Aircraft winged target showing a different arrangement of spoilers (?) on the fuselage. Painting alone on what was after all a disposable asset must have cost a considerable sum.*

32: *First cousin to the bat-eared fox? This is the American Aero 27 winged target. The ribbed wing surfaces are strongly reminiscent of Boeing aircraft of the '30s.*

33: *An equally strange-looking contraption was the RFD winged target, although it at least looks relatively straightforward in construction.*

35: The winged target taken to its logical conclusion. This is a radio-controlled De Havilland Queen Bee, K8652 of 3 AACU, being launched from HMS Shropshire on 6 January 1939 for gunnery practice. The jolly jack tars made sure it was her last flight.

36 Below: Post-war, numerous Fairey Fireflys were converted to radio-controlled targets. This is the prototype U Mk 9, WB257, in April 1956.The wingtip pods held cameras which recorded the flight of missiles approaching the aircraft.

37 Above: Canberra B Mk 2 WK121 arrived at the A&AEE in February 1960 and stayed for 13 years. Here it is at 150 knots at 10 ft (3 m) altitude during a 'snatch' at Lulworth Ranges.

THE AIR SNATCH TECHNIQUE
Low, not slow

Nobody could fail to be thrilled at the sight of an air snatch being carried out. As far as the target was concerned it was just another target, but the method of getting it airborne and landing it again after the sortie was quite the most exciting of all target towing excercises.

The idea was not new as far as the RAF was concerned as they were using a similar method in the early 'thirties when the Army Co-operation Squadrons were using this method for picking up messages from the ground using such aircraft as the Horsley, Audax and Hector. As far as I know, the Seahawk was the first aircraft to use the method for target towing using a Mk l Dart Target on a 4,000 ft (1,220 m) length of nylon tow rope. An 8 ft (2.4 m) pole with a hook on the end was attached to the deck arrester hook. This pole was attached by an electro-mechanical release slip which allowed the pole, rope and target to be jettisoned after the sortie.

The target was placed on the ground with the tow rope attached, the rope being laid out in a straight line behind the target with a large loop at the other end. The loop was supported off the ground by a pair of 12 ft (3.6 m) high poles usually referred to as 'goal posts'. Flying at a speed of around 150 knots, the tow aircraft approached at a height of 10 to 12 ft (3 to 3.6 m). As soon as the tow hook engaged with the loop, the aircraft increased the power

and went into a steep climb, taking the rope with the target on the end which lifted into the air almost vertically.

This proceedure called for a high degree of skill on the part of the pilot and a method of practice was used whereby a bare loop was suspended from the goal post and, as soon as the loop had been snatched by the aircraft, another was fitted and the aircraft could return and snatch the second. Up to four loops could be snatched in this way and carried on the same hook without the rope or target, thus providing the necessary practice for the live run.

The ground party who were responsible for laying out the target were under the direction of a ground controller who was in contact with the pilot by radio. The controller would advise the pilot on his run-in, informing him if he was too high or too low and if the hook was engaged with the rope so that the aircraft could start its climb. The controller would also confirm when the target actually lifted off.

When the aircraft returned after the sortie, the pilot had to drop the hook, rope and target onto the dropping zone. This was difficult as the target was 4,000 ft (1,220 m) behind the aircraft and it was not easy to judge when the target had reached the 'DZ'. Therefore the pilot dropped the target on the directions of the ground controller, which usually went along these lines: *"DZ clear for drop. Surface wind*

240—15 knots. Come left—left—steady. Counting down. Five—four—three—two—one. Now! Target released. We've got it. Controller closing down."

Then would come the hard work. One target to dig out of the ground and four thousand feet of rope to wind in. This could often be hooked up over trees and bushes. It was usually in a tangle.

Very often, the training of new pilots for the snatch routine was a hairy process. I particularly remember one pilot, who eventually turned out to be one of the best, having tremendous difficulty during his first attempts. With a practice loop set up on the posts, he made no less than seven runs, only to be told by the controller after each, *"Sorry. Two feet too high"*. Finally, on the eighth run, in sheer desperation, he came boring across the airfield at about ten feet (3 m) with his hook ploughing a furrow across the grass. Despite my frantic shouts of *"Up! Up! Up!"* he flew straight into the goal posts and collected the loop around the nose of the aircraft. Thankfully there was no target on the end of the loop. He could never have got rid of it.

After landing, and following a nerve-steadying smoke and a couple of brandies, he took off again, did a perfect snatch first time, and to my knowledge, never missed another.

Later on, when the Canberra was being used as a snatch vehicle, I was given the opportunity to fly in the bomb aimers position in the nose during a snatch. Coming across the airfield at 20 ft (6 m), travelling at 140 knots, I think it was the biggest thrill of my life. I remember telling the ground party afterwards, *"If you think the snatch is spectacular, just go up there and try it"*. Spectacular it certainly was, so much so that we were invited to perform snatches at various RAF Stations for their Battle of Britain displays. Over the years Exeter, Coltishall, Colerne, Benson, Chivenor and St. Mawgan were all visited.

These public displays were great fun and the ground party worked out a routine almost like a circus act. It was too hazardous to snatch a dart target with 4,000 ft of rope so we worked out a routine using a 3 x 15 sleeve target on 800 ft (240 m) which worked quite well. After the snatch, the aircraft did a circuit and dropped the target in the middle of the airfield on the next time round.

As each different airfield meant a different approach and surroundings, even the best snatch pilot could be excused for not taking the loop on the first attempt, and it is amusing to recall the system we used to save the pilot any embarrassment if he missed on the first attempt. During the briefing, we would instruct the commentator who would be telling the crowd what was going on over the public address system, to tell them that the pilot usually did a dummy run before coming in for a live snatch. If the pilot was fortunate enough to take it on the first attempt, the commentator would say: *"I see he is not having his usual dummy run this time"*. Fortunately, a third run was never necessary.

To return to the subject of real target towing, it was stated earlier that the first snatches were used in conjunction with the Mk 1 Dart Target. This was a high-speed expendable target consisting of four plywood fins bolted on to a central spine made from angle iron. The target was 12 ft (3.6 m) long and 4 ft (0.6 m) across from fin tip to fin tip. It was fitted with a hinged towing bracket situated at the point of balance or centre of gravity and weighed 130 lbs (59 kg) and was fitted with a sheet metal radar reflector. There was no parachute recovery system and after it was dropped following a sortie, it usually hurtled down and drove itself six feet into the ground — which called for a couple of hours of pick and shovel work to recover the remains.

An slightly improved version of this target was very similar and was referred to as the Mk 2.

Various other types of aircraft were used for the

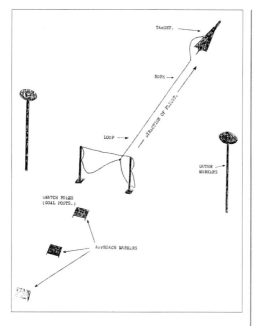

snatch technique, but for the job of general target towing it was principally the Royal Navy who had the requirement. They used Sea Hawk, Scimitar and Sea Vixen aircraft for this purpose. It was quite a simple job with these aircraft to attach the towing pole pole to the existing arrestor hook, but when the Canberra was used as a towing vehicle, a special hook had to be designed. This was fitted with a fin and a horizontal stabiliser to make the hook fly steady and force it to fly downwards. Without the stabiliser the hook would have streamed along close to the fuselage. An electro-mechanical release slip was fitted on the fuselage centre line immediately aft of the bomb doors, the hook being attached by a swivel which allowed it to swing in any direction.

Tow line lengths of 4,000 ft (1,220 m) were usual, but if more length was needed an extra 2,000 ft (610 m) was added, making the total length 6,000 ft. When this extra length was used, 4,000 feet was laid out in the normal way, the remainder being packed into a dispenser which was placed on one side of the snatch line close to the goal posts. It was very impressive during the snatch to see the rope being pulled out of the dispenser at around 2,000 ft per second. As the slightest knot or kink would cause the rope to break, this emphasised the care needed when packing the rope into the dispenser.

Despite the glamour and excitement of the air snatch, it was not used extensively by the Services on account of the elaborate ground facilities needed. Consequently it seemed that the air snatch technique was destined to fade into obscurity. This, however, turned out not to be the case as it was to play a very important part in the armament development programme.

In 1959 the Armament and Aircraft Experimental Establishment (A&AEE) at Boscombe Down were feeling the need for a target vehicle to carry special research equipment. It was one thing to count the number of holes in a target after a shoot to find out how many hits had been scored, but the armament scientists wanted to know what had happened to the ones that had missed. An intelligent target was obviously the best way to find out.

The French had been using the air snatch technique with a Canberra aircraft to fly some enormous targets known as the S100 and the Javelot, both being fitted with a workable near-miss detection system. It was decided to purchase a number of these for armament research at Boscombe Down.

The S100 was a monstrous target which resembled a wartime V2 rocket which had been manufactured by a firm of agricultural engineers. It was 14 feet (4.2 m) long and weighed some 450 lbs (204 kg). The Javelot was similar in size but weighed 100 lbs (45 kg) less and was a much more workmanlike job. Both types were of all metal construction using a central spine with fitted frames around to carry the cowlings. Brackets and boxes were fitted to the spine to carry the near-miss recording gear, the batteries and so on. Both targets were fitted with a parachute recovery system to bring the target down safely after a sortie. A towing stirrup was fitted which hinged around the centre of gravity position on the target and this stirrup was used to operate the parachute recovery system.

When the target was being towed through the air, the towing stirrup was in the forward position, and when the rope was released from the aircraft, the target tended to fly on with the rope dragging behind. This rope drag had the tendency to force the towing stirrup rearwards, in turn causing a pin to be pulled from an electrical switch which had the effect of detonating the explosive bolts which held the rear door in place, this door being used to keep the recovery parachute in place. Removal of the door allowed the parachute to deploy thus arresting the target's forward speed and lowering it to the ground.

In theory, this sounded fine. It meant that the targets with their expensive equipment could be used over and over again. In practice, however, things did not go exactly as planned. Failure after failure occurred. Not with the the the firing sortie and the near-miss recorder for this gave the boffins exactly what they wanted, but the problems were with the parachute recovery system. It can be stated with certainty that an S100 target whistling down without a parachute from several hundred feet up makes a tidy hole in the ground. There is not much left to salvage afterwards.

There were several reasons for the failures. First of all it was found that the batteries which powered the recording gear were also used to fire the explosive bolts. As a target was normally flown until the recording gear ceased to function due to the batteries being exhausted it meant that there was no power left for the explosive bolts. This was easily remedied by fitting separate batteries for the recovery system,

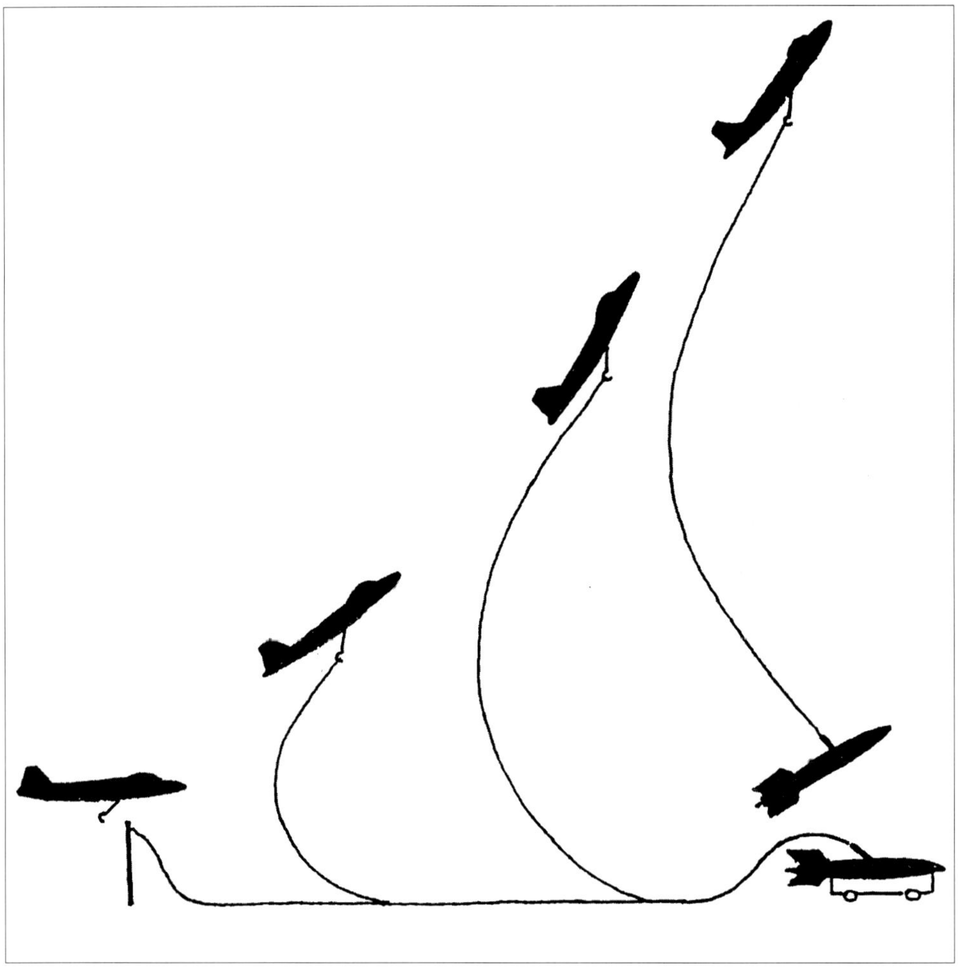

40: *This diagram shows in exaggerated form the track of the aircraft and target during a snatch pickup.*

but there were other problems which were not so easy to cure.

If, for some reason, the towing aircraft decelerated too sharply, there was a tendency for the target to fly on and overtake the aircraft resulting in the rope dragging the towing stirrup rearwards and releasing the parachute. This premature deployment ripped the parachute clean off the target leaving nothing to come home with. The S100 target was particularly prone to swinging like a pendulum from its parachute when descending. If it struck the ground at the wrong end of the swing, the target broke its back. These were some of the problems encountered with the French targets, but, to be fair, when they operated successfully, they flew well and the intelligence they carried was just what the armament kings wanted.

At Boscombe Down, we ran through the first batch of French targets at an alarming rate and the purchase of the second batch was being considered. These targets were expensive, and while it was being argued whether the end justified the means, the Towed Target Development Section at Boscombe thought they could do better. Models and plans were made and with the co-operation of the Squadron Engineer and the Project Pilot, a full sized prototype was built which was successfully flown and recovered from Lulworth Ranges in June, 1960.

The target was named 'Trident' from the three-man team who designed and built it and it proved to be an outstanding success. It flew well and had a very reliable parachute recovery system, indeed, one particular target was a veteran of 23 sorties. It was extremely light in construction weighing only 130 lbs (59 kg) and consisted of a light alloy tube, 12 in (30 cm) in diameter with four light alloy fins rivetted to the rear end. A nose cone which incorporated a pneumatic shock absorber was fitted to the front end of the tubular body, this being removable. Provision was made for the trays which carried the recording gear which slid down into the tubular body and held in position with a pin. The trays could be changed in a matter of minutes simply by removing the nose cone. The recovery parachute was housed in the rear end of the target and the release was effected in the normal way by the rearward movement of the towing stirrup, the actual parachute release being a simple mechanical device. Trident was assembled in three separate sections, front, centre and rear. These sections were interchangable which permitted speedy repair by section replacement.

The Tridents replaced the French targets and were used extensively for armament trials at A&AEE for several years. Twelve were produced by Brooklands Aviation at Northampton at a quarter of the price of the French targets and averaged 20 flights

41: Some of the recovery team demonstrating why the Mk 2 Dart Target was not entirely ideal for its purpose.

42: Canberra B Mk 6 WH952 shows off its snatch hook sometime in the later 1950s.

43: An S100 Target just after it has been 'snatched' off its trolley. Its similarity to the German V2 missile is readily apparent. Dunkeswell 1961.

44: *A French S100 Dart target on an Oxford carrier ready for snatching at Lulworth. The Carrier, Tracked, Oxford Mk 1, to give it its army designation, was a relic of World War II. Too late for that conflict, 400 were produced by Wolseley in 1946 for gun towing and general fetching and carrying, as seen here.*

45: *Preparing a French Javelot target for snatching from the Lulworth Ranges. Baden-Powell would probably have felt quite at home with those knots.*

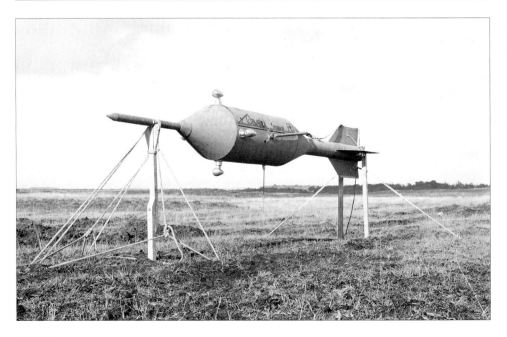

46: *Et voila! The Javelot is ready for the snatch. It is easy to imagine what the state of the target would be shortly after it was released to find its own way back to earth.*

per target, as opposed to the average of three flights for the S100 and five for the Javelot.

During the period that the snatch targets were being operated by the A&AEE, very few were snatched from the airfield at Boscombe Down on account of the fact that they could not be dropped there due to the interference with air traffic. Those targets that were snatched at Boscombe therefore had to be dropped at an Army dropping zone at Everleigh, 12 miles (19 km) away. This was undesirable as the ground party had to leave the airfield and travel to the dropping zone as soon as possible after the snatch. Any hold up or breakdown could mean that the aircraft returned to the DZ before the arrival of the ground crew, leaving the pilot without a talkdown facility.

At first, to ease the situation, the Royal Armoured Corps firing range at Lulworth was used. Although this had the advantage of cutting down the overland towing distance, the trials had to be slotted into the Army's firing programme which was not always convenient. The overland towing distance was an important factor during the armament firing trials. To start with there was always the danger of a rope breakage, and although a comparitively safe route was chosen, there was no telling where a target would land if a breakage occured. A chase aircraft was always used to fly alongside the target when flying overland.

Eventually, the problem was solved when a disused airfield in Devon was taken over which was only four minutes flying time from Lyme Bay where the firing ranges were situated. This also meant that a chase aircraft was not required. Dunkeswell therefore became the home of the Boscombe Down Towed Target Section for the next four years with as many

47 Below: But no! The French know how to build a solid target. This is the Javelot being recovered by means of an Army Comet tank, somewhere on the Lulworth Ranges in 1961.

as four sorties a week taking place. Although this has little to do with the history of the towed target, I make no excuse for making this the longest chapter of all as it was a period when an immense amount of development took place. For the author, involved as I was for many years in aerial targetry, it was the most exciting period of my life.

The Dart targets, the S100, Javelot and finally Trident, showed what type of target was really needed to keep pace with modern weaponry. The snatch technique was too involved for regular Service use and unfortunately the tow length could not be extended beyond 6,000 feet. The answer lay in a similar type of target being air-launched and paid out on a long tow winch. The French did some work in air-launching a Javelot target from a Canberra aircraft with the rope being carried in dispensers in the bomb bay. This of course did away with the snatch, but did nothing to lengthen the tow length and still required a large dropping zone to drop the target after a sortie.

This would appear to close the chapter on the air snatch technique but I always feel that system could always have a use for the initial flight testing of new targets. There is a very good reason for saying this. With the snatch method, once the aircraft has taken the rope, the target is too far behind to affect the tug no matter what erratic behaviour the target may indulge in. This was the main reason why the Trident came into being. It was thought at the time, if it flies, so much the better; if it doesn't — just cut the rope.

It was this safety advantage that led to the snatch method being used for the most extraordinary target of all. This was a requirement from the Royal Ra-

dar Establishment at Malvern, who had a need for a 4 ft (1.2 m) diameter sphere to be towed through the air on 4,000 ft of rope. The sphere was for radar calibration and was made from fibreglass coated with copper. It only weighed 40 lbs (18 kg) and flew extremely well. What other means could have been used to get such a target airborne?

Perhaps we have not seen the last of the air snatch technique, and possibly some day we may again see the spectacle of an aircraft streaking in at near zero feet with its hook poised to snatch some new shape into the sky.

48: Canberra B Mk 2 WD945, still in Medium Sea Grey and Gloss Night Bomber Command camouflage of the period, about to make a snatch, sometime in the mid-1950s.

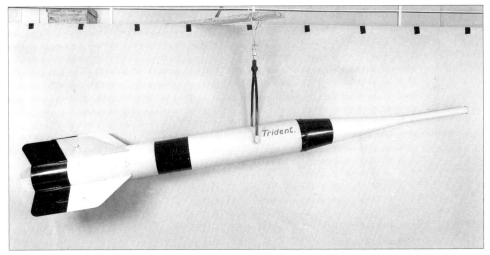

49: The Trident Target, seen here on 22 August 1962, developed under the aegis of the A&AEE, and for which the author was later awarded the British Empire Medal.

50: Another view of the French S100 Target being prepared for a mission. It really should have been turned into a missile.

TARGET INTELLIGENCE
Don't hit the target

In the proceeding chapters we have talked about Target Intelligence and electronic equipment. Before passing on to modern targets it would be as well to examine some of this equipment and the need for its use.

Essentially the subject can be split up into two main groups. First of all there is the acquisition source for the homing missiles, which are normally flares of some description which are activated by radio signals from the towing aircraft. Then there are the radar reflectors; the older type of these were sheet metal corner reflectors such as the Anning reflector. This was used on the banner targets and the parachute recovery doors which normally took the form of a reflector on the Mk 2 Dart Target and most of the snatch targets. There were also the different type of materials with wire mesh woven into the weave as used on banner targets. Later targets are fitted with a Luneberg lens which is a remarkable device taking the form of a small sphere, which shows on the radar screen an image comparable to a full-sized aircraft.

Most of these acquisition sources are comparably simple to fit, but Near Miss Recorders or Miss Distance Indicators (MDI) to use the modern teminology, is very much more complex. Why should MDI be necessary? — after all it could be said that a target is there to be shot at and it only matters if it is hit or missed.

The answer to this question is the increasing accuracy of modern gunnery and missiles. Before World War II the use of drone targets was an economical proposition as to score a hit on one of these targets was quite an occasion. These days, with guns or missiles being trained by radar, laser or heat, it is unusual to miss and even sleeve targets can be expensive to operate if the first shot can end the entertainment for the day. I remember many years ago when towing sleeves for the Royal Navy cruiser HMS *Tiger*, being asked by the gunnery control officer if we wanted it all at once or a bit at a time. He was true to his word as the sleeve was taken apart by four shots, starting at the tail and working forward until it was all gone.

In these conditions it is desirable to aim at a point some distance away from the target so that a classified hit could be at a point several feet away. It is worth noting at this stage that when towing a sleeve target on 6,000 ft (1,829 m) of cable, the radar-directed naval guns had to be watched very carefully by the fire control officer as there was a tendency for the gun to track up the cable and lock on to the towing aircraft (!).

As far as the British services were concerned, Near Miss Recording was first made possible by the use of the Swedish SAAB system. Used at first in conjunction with the 3 x 15 sleeve target and later on the Brooklands Mk 3 Dart Target, the SAAB system was of the accoustic type using a microphone on the target and a recorder or 'bang box' situated in the winch operator's cockpit. The Type 'G' winch was modified to suit the system by having slip rings fitted to the winch drum shaft and a special 15 cwt (762 kg) breaking strain steel tow cable with an insulated heart strand running through the centre. When a shot was fired close to the target, the microphone picked up the sound of the blast and sent a signal to the recorder via the cable and slip rings. Depending on the strength of the signal picked up by the microphone, the recorder would show a hit on band one, two or three. Each band represented a given distance from the target.

The SAAB system was in use for several years, particularly with the Royal Navy who used it for surface to air gunnery. Unfortunately, the system could only be used with the special conducting tow cable which could not exceed 6,000 ft (1,829 m), this being the maximum length accomodated by the winch.

In the previous chapter, mention was made of the S100, Javelot and Trident targets used by the A&AEE at Boscombe Down for the purpose of air-to-air firing trials. These targets were used to carry the French SFENA near-miss recording system. This was also of the acoustic type but unlike the SAAB sys-

51 Right: A sketch showing the salient points of the Swedish SAAB Miss Distance Indicator and the French SFENA system.

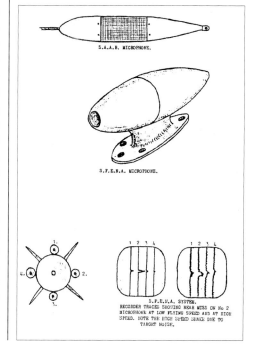

tem, operated independently of the towing aircraft. Four separate microphones were used, these being positioned radially around the target at 90° and were connected to a radio transmitter through an amplifier, each microphone operating on a different channel. The signals were picked up by a ground receiving station which could be as far as 90 miles (145 km) away. The incoming signals were displayed on a four-channel pen recorder. The target was attacked from astern and the noise of the passing shots was picked up by the nearest of the microphones, the recorder then showing the proximity of the shot and whether it was left, right, above or below the target.

The SFENA system was exremely effective but was expensive and inclined to be temperamental. It was not therefore well suited to general target practice. One of its problems was the air noise of the target which increased with a corresponding increase in speed. This noise was then picked up by the microphones. At speeds in excess of 350 knots the air noise of the S100 and Javelot targets made it difficult to identify the shots on the recorder. The SAAB system did not suffer from this problem as the microphones were much less sensitive, being designed to detect the noise of bursting shells instead of the sound of passing shots as was the case of the SFENA. Acoustic systems have since been developed with selective microphones which can filter out all but the required noises.

Another very interesting MDI system was produced by ML Aviation of White Waltham, specifically for the A&AEE firing trials. This was the ARAS (Aerial Rocket Assessment System) which was used on firing trials with the Lightning and Sea Vixen aircraft using 2 inch (5 cm) rockets. The ARAS used radio waves for its operation and the Trident used for the trials carried a large tubular aerial attached to the aft end which transmitted a beam radially around the target. Attacks were made from astern and the rockets, which were fitted with detector heads, exploded with a bright flash as they passed through the radio beam. The flashes of the exploding rounds were recorded on cameras on the attacking aircraft.

The ARAS worked extremely well and played an important part in the development of the 2 inch rocket. Some work was carried out in producing detector heads for 30 mm cannon shells but there is no record of these being actually used. Like the SFENA the ARAS was expensive to operate and while its use was justified in weapons research, it was hardly feasible for every day target practice.

The Del Mar Engineering Laboratories of Los Angeles, USA, have been in the forefront of target system development for many years and have produced the Acoustiscore system which is in use with the American Services. This system uses a single microphone and is in many ways similar to the SFENA system, consisting basically of a sensor or microphone, a signalling conditioning unit, a transmission

unit and a data reception and display unit. The sensor is attached externally to the target and is housed in an enclosure which is designed to eliminate spurious signals made by the air noise of the target. Data from the target is transmitted to the ground station and the system is capable of handling scoring rates of up to 6,000 rounds per minute from aerial targets travelling up to supersonic speeds.

The American system is much later than the SFENA and is therefore much more advanced in design, being less temperamental, more positive in operation and less expensive to use, all of which makes it a much more feasible system for day to day firing practice.

Another system which has been used by the Americans and may be considered for use in this country is the Franklin RAMDI (Radio Active Miss Distance Indication.) This is a type of Geiger counter in the target and the shells use a radioactive substance in the head. As the projectile passes the target, the meter picks up the presence of the radioactive source and transmits the information to the towing aircraft or ground station. The radioactive substance is in powder form and is housed in a magnesium container. As the spent projectile falls into the sea after firing, the magnesium dissolves in the salt water allowing the powder to dissipate, thus rendering the source harmless. This type of system is obviously restricted to sea firing ranges.

The Delmar (as usually written) Company was also responsible for another interesting device. This was a tracking system which was activated by the towing aircraft. The target ejected electrically fired smoke puffs at regular intervals with a most spectacular result. A dotted line appeared in the sky at the head of

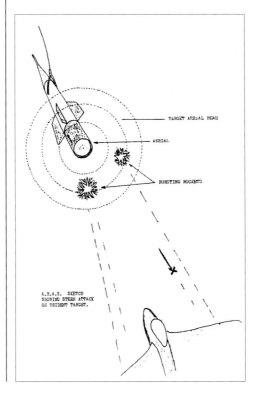

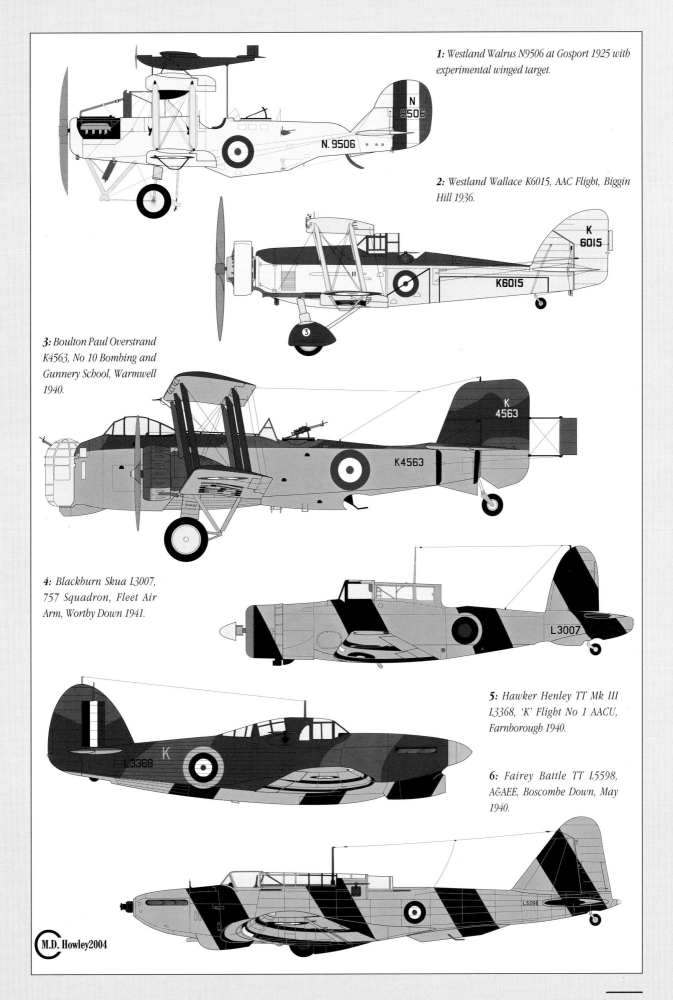

1: Westland Walrus N9506 at Gosport 1925 with experimental winged target.

2: Westland Wallace K6015, AAC Flight, Biggin Hill 1936.

3: Boulton Paul Overstrand K4563, No 10 Bombing and Gunnery School, Warmwell 1940.

4: Blackburn Skua L3007, 757 Squadron, Fleet Air Arm, Worthy Down 1941.

5: Hawker Henley TT Mk III L3368, 'K' Flight No 1 AACU, Farnborough 1940.

6: Fairey Battle TT L5598, A&AEE, Boscombe Down, May 1940.

M.D. Howley 2004

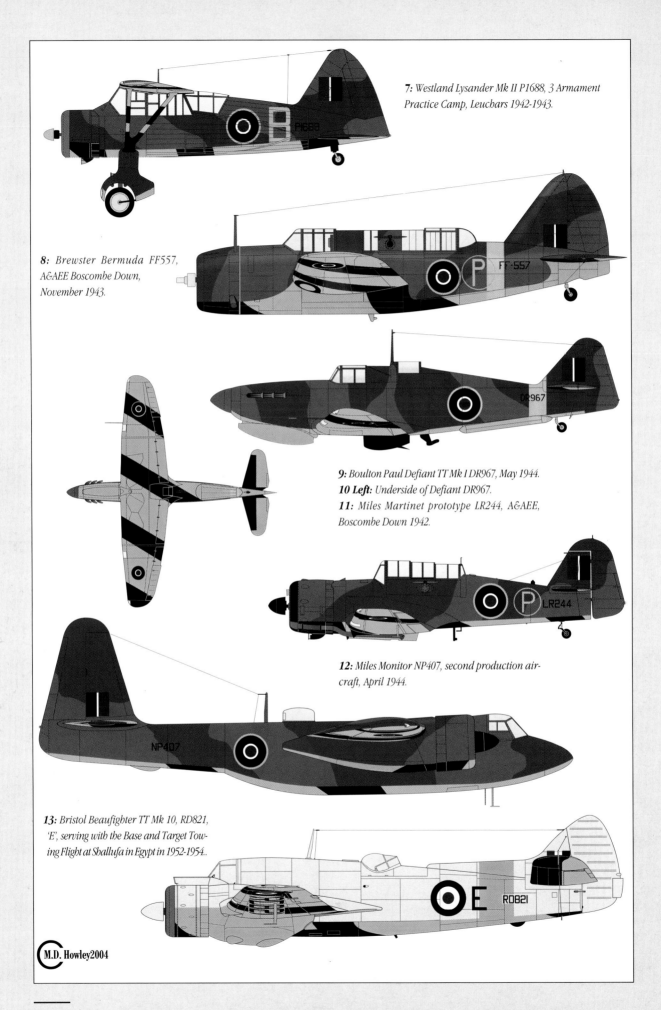

7: *Westland Lysander Mk II P1688, 3 Armament Practice Camp, Leuchars 1942-1943.*

8: *Brewster Bermuda FF557, A&AEE Boscombe Down, November 1943.*

9: *Boulton Paul Defiant TT Mk I DR967, May 1944.*
10 Left: *Underside of Defiant DR967.*
11: *Miles Martinet prototype LR244, A&AEE, Boscombe Down 1942.*

12: *Miles Monitor NP407, second production aircraft, April 1944.*

13: *Bristol Beaufighter TT Mk 10, RD821, 'E', serving with the Base and Target Towing Flight at Shallufa in Egypt in 1952-1954..*

M.D.Howley2004

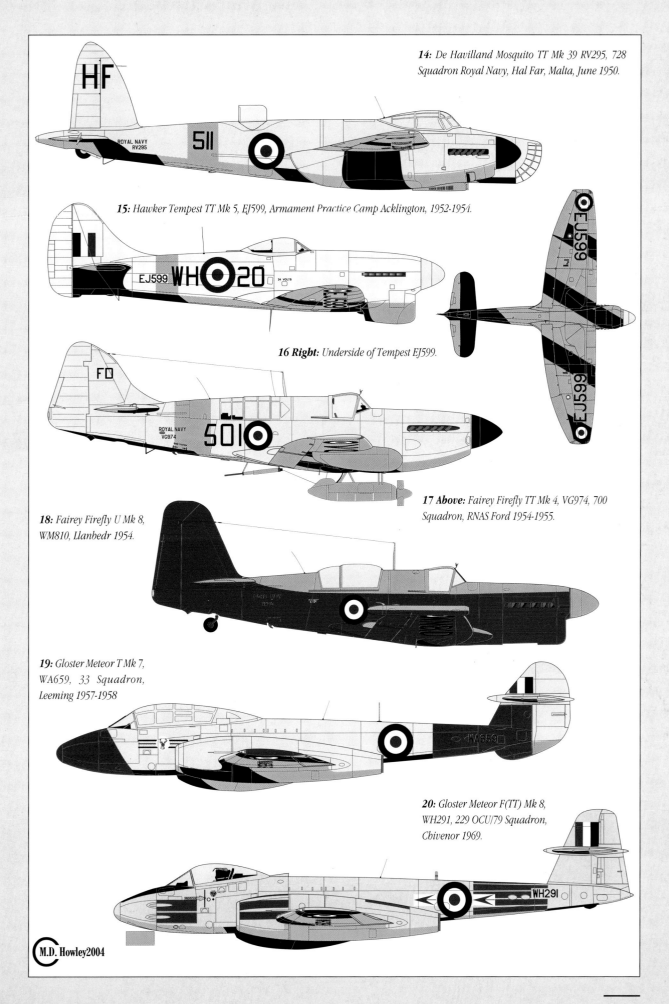

14: *De Havilland Mosquito TT Mk 39 RV295, 728 Squadron Royal Navy, Hal Far, Malta, June 1950.*

15: *Hawker Tempest TT Mk 5, EJ599, Armament Practice Camp Acklington, 1952-1954.*

16 Right: *Underside of Tempest EJ599.*

17 Above: *Fairey Firefly TT Mk 4, VG974, 700 Squadron, RNAS Ford 1954-1955.*

18: *Fairey Firefly U Mk 8, WM810, Llanbedr 1954.*

19: *Gloster Meteor T Mk 7, WA659, 33 Squadron, Leeming 1957-1958*

20: *Gloster Meteor F(TT) Mk 8, WH291, 229 OCU/79 Squadron, Chivenor 1969.*

M.D. Howley 2004

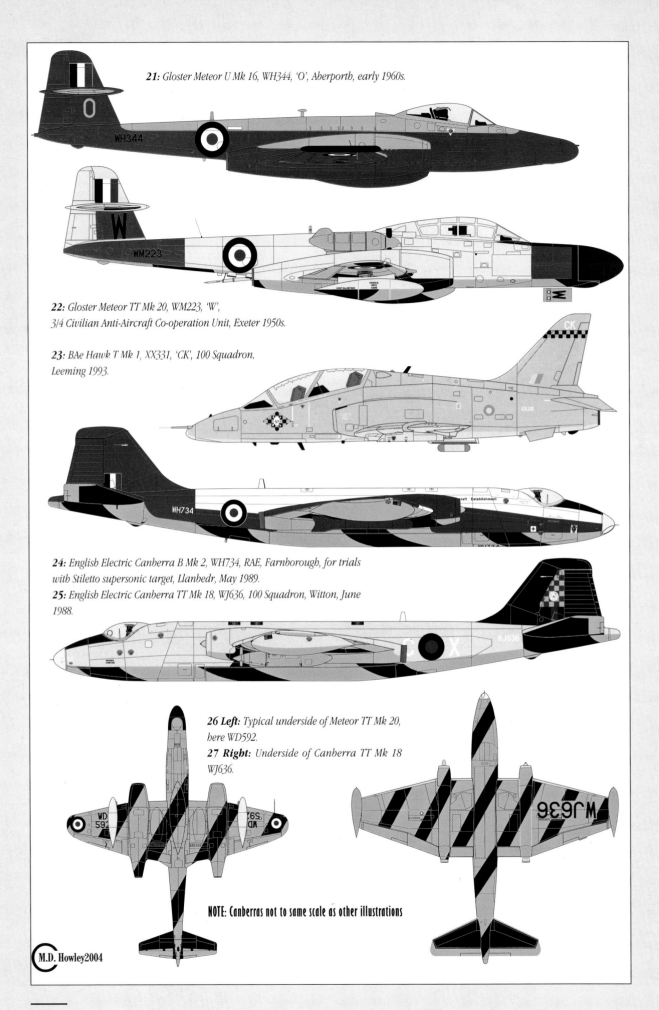

21: *Gloster Meteor U Mk 16, WH344, 'O', Aberporth, early 1960s.*

22: *Gloster Meteor TT Mk 20, WM223, 'W',*
3/4 Civilian Anti-Aircraft Co-operation Unit, Exeter 1950s.

23: *BAe Hawk T Mk 1, XX331, 'CK', 100 Squadron,*
Leeming 1993.

24: *English Electric Canberra B Mk 2, WH734, RAE, Farnborough, for trials*
with Stiletto supersonic target, Llanbedr, May 1989.
25: *English Electric Canberra TT Mk 18, WJ636, 100 Squadron, Witton, June*
1988.

26 Left: *Typical underside of Meteor TT Mk 20,*
here WD592.
27 Right: *Underside of Canberra TT Mk 18*
WJ636.

NOTE: Canberras not to same scale as other illustrations

M.D. Howley 2004

which was the target. This system was used entirely for the identification of the target.

In previous chapters the importance of a realistic target has been stressed. Modern weaponry depends on it. Today, however, the visual appearance is no longer of any importance. It is more important to be realistic on the technical side in such details as radar reflectivity, the heat or light source, the speed and height of the target. The target intelligence systems described in this chapter have now been su-

perseded by later, more effective and reliable equipment. At the end of the day, all of the electronic wizardry still has to be carried through the air, and to do so the modern target has evolved.

53: This is a Mk 3 Dart Target and clearly shows how it was both simple and sophisticated in construction. Comparison with photograph 41 shows that it differed considerably from the Mk 2, these differences being exaggerated by the perspective here which makes it look very short.

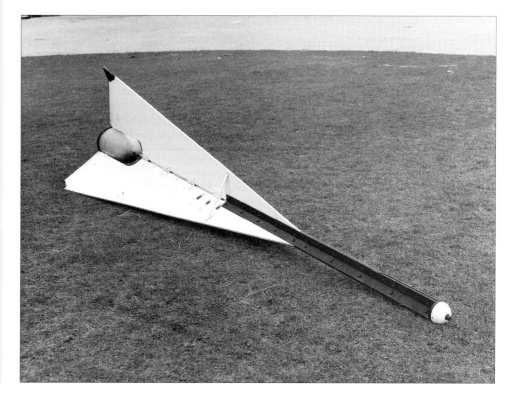

54: A front view of the Mk 3 Dart Target. In this picture its substantial size, some 12 ft (3.6 m) long, is not readily apparent, but its dart-like proportions are clear.

MODERN TARGETS
Size is not important

If we think of modern targets as those which were built to carry the intelligence referred to in the previous chapter, then we must first think of the Mk 3 Dart Target. Operation of the Mk 1 and 2 Dart Targets showed that dart targets could, by virtue of their low drag, be operated at higher speeds than banners or sleeves and were extremely good for damage absorbtion. The great disadvantage with the dart was the complication of the air snatch technique.

It was therefore logical to think in terms of an air-launched dart target which could be paid out on a winch. The result of this thinking was the Brooklands Mk 3 Dart Target. Designed for use with the Meteor TT Mk 20 using a 'G' Type winch, the Mk 3 was able to make use of the SAAB Near Miss Recording System which had previously been used on sleeve targets.

The Mk 3 was 15 ft (4.5 m) long and 4 ft (1.2m) wide at its maximum width. It weighed 180 lbs (82 kg) and consisted of three metal-covered plywood fins mounted on a light allow spine. The lower fin was hinged allowing it to be folded upwards to allow sufficient ground clearance when stowed on the aircraft. At the rear end of the target, mounted centrally between the fins was a fibreglass parachute container, the front end of this being shaped to fit the SAAB microphone. The cover plate of the parachute container was fitted with four vanes which acted as a corner radar reflector, thereby making the target radar reflective in all planes. A hinged towing bracket

was fitted at the centre of gravity point, which was connected to the 25 ft (7.5 m) halyard, in turn stowed in a dispenser immediately forward of the towing bracket. The other end of the halyard was attached to the winch cable. The halyard was of 3,000 lb (1,360 kg) breaking strain nylon with an insulated heart strand to provide a conductor for the microphone.

The target was fitted to the underside of the aircraft fuselage and held in position by an electro-mechanical release hook. The nylon halyard was attached to the winch cable by a special release hook and link. This link was pulled firmly on to a rearwards facing hook on the underside of the fuselage by the winch cable and the winch brake applied.

When the target release switch was operated, a rotary solenoid released the catch allowing the lower fin to swing down and lock into place. A time delay switch operated three seconds later allowing the target to drop away, pulling the nylon halyard from its dispenser as it fell. This left the target flying under the aircraft with the end of the halyard still held in position suspended from the rearwards facing hook. The winch was then selected to pay out and as the tension was released on the cable, the link slid back off the rearwards facing hook allowing the target to fly freely on the main towing cable.

After the sortie the target was winched in and, when the joining link reached the buffer strut on the aircraft, the scissors type release opened up allowing the halyard to fall away with the target. As the target fell away, the airstream dragged the halyard to

55 Above: *Meteor TT Mk 20 WM234 was originally an NF Mk 11, serving with 264 and 68 Squadrons before conversion. Here it is in the mid-1960s fitted with a Mk 3 launcher and a Rushton Mk 2 target. Escalating costs associated with the use of drone targets as guided missile technology improved led to a demand for new towed targets. WM234 was used extensively by Flight Refuelling Ltd during trials of the American designed Hayes T.6 and T.17 targets and Del Mar winch, which were intended for licence production. In the event Flight Refuelling's own Rushton target and winch were preferred by the Ministry. Finished in a combination of silver, black, yellow and day-glo panels, the aircraft was scrapped in July 1970. Parts still survive in various museums.*

the rear, pulling the towing arm which released the parachute rear door allowing the parachute to lower the target safely to the ground.

The forerunners of the modem concept of targets in this country were the snatch targets described in the previous chapter. No more mention of these is needed except to say that they were the first to use the 'black box in a tube' principle in use today.

One of the first of the new 'space age' targets was the Del Mar which was used in conjunction with the Del Mar DX4 winch (or tow reel to use the American term). This very simple but beautifully engineered little winch was capable of paying out 25,000 ft (7,620 m) of piano wire cable with a thickness of only 0.050 inches (1.27 mm) in diameter.

The Del Mar Target was shaped like a fat aerial bomb, 5 ft 6 in (1.65 m) in length and 2 ft 6 in (0.76 m) in diameter. It was extremely light weighing only 11 lbs (5 kg) when ready for flight. It was constructed from $^1/_8$ in (3 mm) thick hardboard which was moulded to shape and the four fins were made from expanded polystyrene. The target body was held in shape by polystyrene bulkheads covered with aluminium foil which acted as a radar reflector operating in all planes. Attachment to the cable was by a small swivel in the nose which allowed the target to rotate around its longtitudinal axis. The polystyrene fins were so shaped that the rear 6 in (150 mm) were offset 15°. As the target was towed through the air, it rotated at high speed which increased its stability.

When stowed on the aircraft, the target was housed in a rearwards facing metal basket from which the target was launched and paid out. After the sortie, the target could be recovered by winching it back into the basket. If the target was damaged and unstable it was difficult to effect a recovery. In this case, an electrically operated explosive cable cutter was used to jettison it.

The Del Mar was a very cheap and effective target and although it had the advantage of a long tow length, it could not be operated at speeds in excess of 350 knots as the fragile angled fin tips tended to snap off. Although a radio-activated infra-red flare could be fitted, due to its high rotational speed it was not practical to fit any kind of miss distance indi-

cation. After trials at Boscombe Down, it was decided that it was not quite what the Services wanted.

Despite the drawbacks with the Del Mar system, it was found that the winch was a winner and it was decided to fit one to the Meteor TT Mk 20 for use with the new Rushton Target which Flight Refuelling Ltd were intending to produce after difficulties with licence-building the American Hayes target.

This was a period when some pilots and gunners were heard to criticise some targets as being too small and difficult to see. If this criticism had been well founded then the Rushton would have been useless, as it was quite the smallest target to date. With the modern weapons guidance systems now in use, however, size is no longer of any importance as long as the target can carry its intelligence and homing source. This the Rushton can do very well and its small size, light weight and clean lines give the target a low drag enabling higher towing speeds to be used.

Essentially, the Rushton Target is a 7 in (180 mm) diameter tube, approximately 10 ft (3 m) long, with a rounded nose and tail and four triangular shaped fins attached at the rear. Its weight varies with the equipment carried but it is usually between 100 and 140 lbs (45 to 63.5 kg). Both nose and tail are removable to give access to the equipment carried inside, and these removable portions usually take the form of a Luneberg lens, devices which magnify the radar image. Each of the tail fins consist of two metal skins with a 2 in (50 mm) gap in between, but angled together in the front to form the leading edge. The gap at the trailing edge is used to house the infra-red flares which the target carries. The target is towed by a swivel attachment at the centre of gravity position and on the TT Mk 20 aircraft it was carried on a launcher fitted to the underside of the aircraft. This launcher allowed the target to be released, paid out, hauled in and recovered back to the aircraft. It is also fitted with an electrically-operated explosive cable cutter for emergency use. The Meteor/Del Mar/Rushton link-up proved to be a highly successful combination; so much so that the Navy decided to try a similar system on the Sea Vixen. Trials were subsequently carried out at Boscombe Down. In this case, the winch was carried under the port mainplane on

56: *This publicity shot from Flight Refuelling Ltd, the predecessor of Cobham Plc, says it all. A Rushton towed target with 10 in (25 cm) forward and 7 in (18 cm) rearward facing Luneberg lens. The small size of the target is apparent from the black and white 1 foot (30 cm) scale bar on the stand placard.*

the standard NATO stores pylon, while the target launcher was mounted outboard of the winch. Unfortunately, the operator's window was on the starboard side of the aircraft, but for some obscure technical reason, the target system could not be mounted on that side. This meant that the operator had to watch the winch and target through a periscope. Even so, the system proved quite satisfactory and the asymmetric tow did not present any problems. In the event, the Sea Vixen did not enter service as a target tower, probably due to the imminent arrival of the Canberra TT Mk 18.

For many years this was the standard target tower in service with the RAF and Royal Navy, the last not being retired until 1997 when age finally began to catch up with the airframe and engines*. It made use of the sophisticated Rushton Winch which is capable of carrying 50,000 ft (22,680 m) of towing cable. It incorporated the launcher for the Rushton Target which has not altered greatly since its initial use with the Meteor TT Mk 20.

At the time of writing, apart from the unmanned drone targets, most of the target towing capability of the Royal Air Force now rests with the Hawks of 100 Squadron which can carry Rushton targets on fuselage belly racks with the minimum of modification.

There are many variations of the Rushton Target, including a sea-skimmer which can fly at a set distance from the surface to simulate an Exocet-type of missile. This gives what was always wanted — a truly realistic target. Many other different types of modern target are in use in various parts of the world, most of which are similar in shape and size to the Rushton. With world target design funneling down to the same basic shape, we would appear to be approaching the ultimate. But what about targets being towed at Mach 2 and 3? If our targets are to be realistic, we shall need them.

*This was not the end of the Canberra in RAF service, however, for in June 2004 No 1 PRU (39 Squadron) still retained a dedicated photo-reconnaissance capability with its Canberra PR Mk 9s.

57: Canberra TT Mk 18 WJ682 shows off its 'wasp' target towing markings and the Rushton winch under each wing.

58: A close view of the Rushton winch fitted under the starboard wing of Canberra TT Mk 18, WH718. The winch carries a Rushton target fitted with a Luneberg lens in the nose and what appear to be a cluster of flares on the tail fin. Presumably these would have been connected with missile firing tests.

59: *Another Canberra TT Mk 18, this time WK143, carrying two high-speed targets in piggy-back fashion on the port Rushton winch during trials in 1977-1978..*

60: *This bulbous device is the Del Mar rotating target, pictured during trials at the A&AEE in October 1962. Interestingly marked Meteor TT Mk 20 WD706 sits in the background.*

61 Below: *Canberra TT Mk 18 WK143 releasing a target. Note the fuselage belly modifications carried out earlier by Flight Refuelling in connection with development of their Mk 20 refuelling pod.*

WINCHES
Letting it all hang out

The airborne towed target winch was introduced when it became necessary to replace the rope or cord because of the longer tow lengths required. Rope suffers from enormous drag when towed through the air at speed and this drag imposes greater loads on the rope. As stated earlier in this survey, 8,000 ft (2,438 m) of $^7/_{16}$ inch (13 mm) diameter nylon cord will give 3,000 lbs (1,360 kg) of drag, which is equal to the breaking strain of the rope. Flexible steel cable, however, being very much smaller in diameter and having a smoother finish gives a very much reduced drag.

Rope can be folded in to a dispenser in a figure-of-eight pattern and several thousand feet of rope can be stowed this way and still allow a free 'runout' when the target is released. Steel cable, however, cannot be packed in this way as it would kink and twist resulting in subsequent breakage. The obvious answer was to wrap the cable around a drum which could rotate and and allow a free runout of the cable. The other clear advantage was that if the drum

was power driven, the cable could be reeled back in again after use.

One of the first types to be used was the American 'Jupiter' Aero Tow Reel which came into use in the late 'thirties. The Jupiter carried 1,200 ft (366 m) of 10 cwt (1120 lbs/508 kg) breaking strain cable on a free running drum. There was no power drive to the drum, but a brake was fitted which could be controlled from the cockpit. The target, usually a sleeve, was attached to the end of the cable and stowed on the aircraft. When ready to stream, the target was released with the winch brake on. As the target was streaming the brake was released, allowing the cable to run off the drum, pulled by the drag of the target.

The cable continued to run until the last layer on the drum was exposed. At this stage, a lever which had been held in position by the wrapping cable was allowed to spring out, automatically applying the brake, thus leaving the target flying on the maximum tow length. On completion of the firing sortie, the brake was again released from the cockpit allowing

62 **Above:** *This very clear picture of the winch operator's position in the Henley shows both the projecting arm of the Type 'B' winch, the cable drum and the windmill in the horizontal position. When in use it was rotated through 90° to face into the air stream. The placard next to the circular revolution counter shows the length of cable tow by reference to the small counter to the right. This shows the length of cable paid out in feet and the equivalent number showing on the counter, as follows:*

Cable length	Counter
1,000	550
2,000	830
3,000	1750
4,000	2410
5,000	3130
6,000	3930
7,000	4830

the drum to rotate and unwind the last few feet of cable which slipped off the drum and allowed the target to fall away. A picture shown in an American magazine of the time showed a Vought SB2U Chesapeake aircraft carrying two Jupiter installations.

Another primitive winch was the hand operated type which was used extensively by Bomber Command for the purpose of training rear gunners. It was normally bolted to the floor of a bomber aircraft just forward of the position of the rear turret, which was removed. A 2 x 10 sleeve target was winched out of the back on a 500 ft (150 m) length of 10 cwt (508 kg) cable and the mid-upper gunner used it for target practice. The target was winched out and in again by hand, with two men using a winding handle on either side. A brake was fitted so that the cable could be paid out under control with the winding handles removed.

The winch itself consisted of a cable drum and a brake drum assembly fitted in a metal frame bolted to the aircraft floor. As the work of winching in called for considerable exertion, the aircraft speed had to be reduced when recovering the target. The hand winch was widely used during World War II and most of the Bomber Command gunners were trained with this system

Although these simple winches and reels are interesting enough, they can hardly compare with the specialised target towing aircraft of today with the sophisticated winches which cater for tow lengths of up to 60,000 ft. It is these and earlier types we shall examine in some detail to illustrate the development of the aerial towed target winch over the past sixty years and more.

Winch Type 'B'

This was probably the most widely used winch of all time, coming in to service before World War II and yet is still being used by some of the smaller foreign air forces today.

The Type 'B' consisted principally of a cable drum driven by a four bladed windmill which was exposed to the airstream. The drum was supported in a rectangular frame mounted in the operator's cockpit of the aircraft with the windmill arm extending through the side of the fuselage. A windmill speed control and a drum brake were provided by a worm and segment mechanism mounted on the windmill arm, both controls being operated from the cockpit. The windmill propeller was mounted on the windmill head which was attached to the windmill arm. Rotation of the arm therefore changed the angle of the blades relative to the slipstream thereby varying the speed of the drive to the cable drum.

When paying the cable out, the windmill was declutched from the cable drum allowing the cable to be hauled out by the drag of the target, the speed being controlled by the brake which was adjusted by the brake control handwheel. To haul in, the windmill was turned to face the slipstream. This action automatically released the brake and engaged the clutch, thus transmitting the drive to the cable drum. The speed was variable over a wide range by use of the windmill control which changed the angle of attack of the windmill and therefore its speed. When winching in, the brake was only used to slow and finally stop the winch when hauling in the last few feet of cable.

The drum contained 7,000 ft (2,134 m) of 10 cwt cable which was wound on the drum evenly by the laying on gear, and the winch was capable of hauling in a 2 x 10 sleeve target at a rate of 1,000 ft (305 m) per minute. Total weight of the winch complete with cable was 340 lbs (154 kg).

In most aircraft, the operator sat facing forward with the winch in front of him, operating it by the use of the brake and windmill controls. The brake was applied by turning the handwheel in a clockwise direction, while the combined clutch and windmill speed control was operated by a cranked handle. During the first part of the movement the windmill was automatically freed from its small band brake and further movement engaged the clutch which was con-

63: A diagram from the Air Publication on the Type 'B' winch showing its main features.

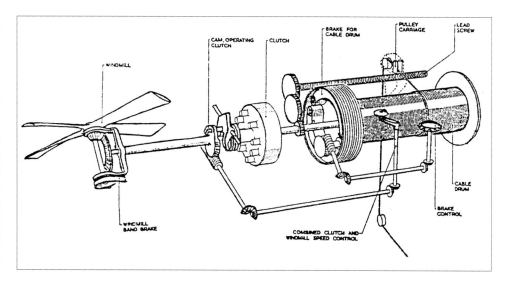

nected to the cable drum. A speed indicator, a footage indicator and a revolution counter was also provided.

In a typical installation such as the Martinet, the cable was fed out of the aircraft over a pulley in the floor and the cable end was pulled up into the cockpit to attach the rigging lines of the sleeve.

The folded target was then pushed out of the aircraft through a hole in the floor and into the slipstream. Once the target was streaming the winch was set to pay out. When hauling in, the cable end had to be hauled right up into the cockpit to release the sleeve. To assist the operator, the last 100 ft (30 m) of cable was painted: red from 100 to 20 ft and yellow for the last 20 ft (6 m). If the end of the cable had been shot away, the operator had no idea of how much cable was left trailing, the footage indicator not being particularly reliable. In such cases it was not unknown for the end of the cable to come thrashing into the cockpit causing injury to the operator. I heard of at least one fatality caused in this way.

The 'B' winch was far from being perfect, but it had a long and distinguished career, dragging more targets than any other. It dragged sleeves, banners, winged targets and anything else that came along. It was fitted to the Hawker Audax and Henley, Fairey Seal, Skua and Battle, Defiant, Martinet, Vengeance, Beaufighter and Lysander aircraft to name just a few. There were many others, some of which are shown in the accompanying photographs.

Winch Type 'D' and 'E'

The Type 'D' winch was a simple frame carrying three drums of cable which were used one after the other. It was similar in use to the American 'Jupiter' winch described earlier in this chapter; once a cable was paid out there was no way of hauling it back in again.

The Type 'E' was basically the same, but was fitted with a 0.4 horsepower electric motor which was not powerful enough to winch in a target, but could manage to winch in a bare cable. The reason for this development was to make use of a new type of release gear which was introduced.

Targets were fitted to the cable in the normal way, but the connection was made using the new release.

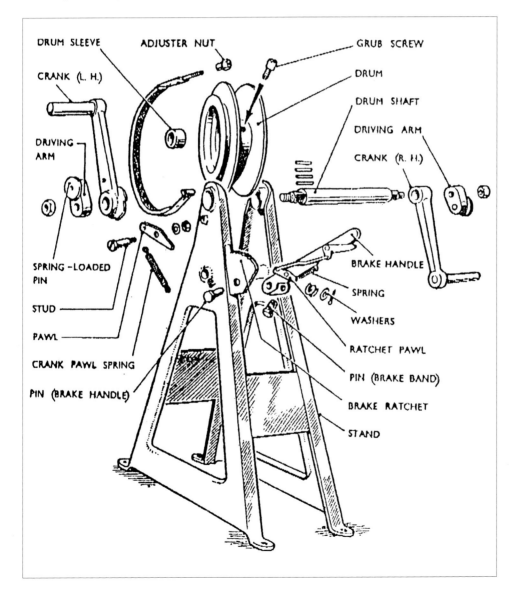

64: Another extract from the official manual, this time detailing the hand cranked Type 'D' winch. It can be imagined what a laborious process hauling in a cable and fitting a new target while in flight would be.

The target was released from the aircraft and the winch set to pay out. When the target was damaged and needed replacing, a metal block or messenger was sent down the cable striking the release gear and slipping the target off the end. The empty cable was then winched in and a new target fitted. If the cable was damaged, preventing the messenger sliding down, it could be slipped from the drum and the second drum used, followed by the third if necessary.

The winch was fitted with an internal expanding brake to control the speed when paying out and a tumbler switch to energise the motor for winding in. When the winding in switch was in the 'off' or 'stop' position, a push button switch could be used to inch in the cable for the last 100 ft (30 m). No automatic laying-on gear was provided but a hand lever was fitted to guide the cable tidily on to the drum.

Winch Type 'F'

The Type 'F' was a further development of the old type 'D'. It was fitted with an 0.5 horsepower electric motor and was provided with an automatic laying-on gear for the cable. On the type 'E' it had been found that the third drum was seldom, if ever, used so the Type 'F' used two drums only.

Another improvement was the control system which employed a single lever to control all operations of the winch. The 'One-Hand Control', as it was known, activated the motor or applied the brake respectively by the forward or backward movement of the lever.

The Types 'D', 'E' and 'F' series were fitted in Martinet, Defiant, Henley and Battle aircraft in the same position in the operators cockpit as the 'B' winch. Although it had none of the power of the 'B' winch, they were simple to operate and did away with the drag of the windmill which allowed slightly higher towing speeds to be used.

Winch Type 'G'

Introduced shortly after World War II, the 'G' Winch was the forerunner of the modern air target winch, and has been widely used by many Air Forces throughout the world, many still being in use today.

The Type 'G' was introduced for use with higher speed towing aircraft such as the Firefly, Tempest, Beaufighter, Mosquito and Meteor, operating sleeve targets and also the winged targets which were coming in to use. It was certainly the first automatic winch and could be operated remotely from the cockpit, being able to pay out and haul in targets at a variable and controllable speed with a maximum winching speed of 1,500 feet (457 m) per minute.

The winch was driven by a two bladed windmill, the pitch of which could be altered while the winch was in operation to vary the speed of winding. The pitch change operated in either direction from the centre or feathered position allowing the winch to

be driven in or out. The Tempest was the only single engined aircraft to use the winch as it was found that it was too much for the pilot to handle on his own and it was more suitable for two seat aircraft where a winch operator could be used.

The 'G' Winch carried 6,300 ft (1,920 m) of 15 cwt (1,680 lb/762 kg) steel cable and would normally pay out 6,100 ft (1,860 m) of this length, the remaining 200 ft (61 m), was left on the drum to secure an anchorage.

When hauling in, the winch automatically stopped with 200 ft of cable left trailing, at this point the 'in limit switch' feathered the propeller which automatically applied the brake. The operator could then press the 'in override switch' which gave the propeller a few degrees of pitch and released the brake. The last 200 ft of cable was then hauled in at a very reduced speed to prevent possible damage due to overwinding.

The operator could tell how much cable was trailing at any given time by the use of an accurate footage counter which consisted of a 12 inch (30 cm) circumference pulley over which the cable ran as it left or returned to the winch. A cam was fitted to the pulley and at each revolution this cam tripped a micro switch sending an impulse to the footage counter mounted on the operators panel. The pulley reversed the operation when hauling in so that the footage counter indicated cable in or out. So accurate was this type of counter that a similar design was used on all subsequent types of winches. The winch itself consisted of a streamlined container divided by bulkheads into three main compartments: the front, which housed the gearbox; the centre containing the cable drum and the tail cone, which carried the brake and pitch change mechanism. The laying-on gear which laid the cable evenly on the drum consisted of a double threaded lead screw laying parallel to the cable drum which carried the traversing carriage forward and aft along the length of the drum. Another screw, laying parallel to the lead screw but with a much finer pitch, carried a nut, which at the extreme ends of the screw operated the in and out limit switches.

The cable was threaded out of the aircraft through a spring loaded telescopic tube known as a buffer strut and a target exchanger was fitted to the end of the cable. When hauling in, the winch would stop as previously stated with 200 ft of cable left trailing. When the in override switch was operated, the remaining cable would come in until the exchanger unit pulled up onto the buffer unit. The buffer strut compressed against the spring until a micro switch was operated. This switch feathered the propeller and finally applied the brake.

The last type of aircraft in the Services to use the 'G' winch was the Meteor TT Mk 20 which presented problems. Previous types of aircraft using the type 'G' carried the winch on the underside of the aircraft allowing the cable to run straight out into the buffer strut. In the case of the TT Mk 20, however, there was insufficient ground clearance for the winch to be underslung in the orthodox manner. It was therefore mounted on a pylon fitted to the upper surface of the starboard mainplane, inboard of the engine nacelles. This necessitated a system of guide tubes to carry the cable to the buffer strut.

In the early 1960s, there was a requirement for a tow length of 60,000 ft (18,288 m) which was considered to be a safe distance when using guided weapons. This could only be achieved by the use of a very thin single strand cable. An attempt was made to convert the aging 'G' winch for this purpose, and modifications including a redesigned gear box to double the speed of the cable drum and different gearing for the laying on gear. The old two bladed propeller was replaced by a new high speed air turbine.

Known as the 'G' Mk 5, the new venture met with moderate success and the fact that a twenty-year old piece of equipment could be modified to do twice the amount of work at twice the speed proved what a fine winch the type 'G' had been and much credit was due to its manufacturers, ML Aviation.

Winch Type 'H'

The Type 'H' was a very advanced design but was not blessed with a particularly good record of servicablity. It was designed mainly for use with what were then high-speed towing aircraft such as the Mosquito TT Mk 19 and Sturgeon TT Mk 3.

The winch was driven by a VSG hydraulic motor which used pressure from three hydraulic pumps which were driven by a windmill. The windmill was fitted in the bomb bay of the aircraft and could be retracted when not in use. In fact, the system was a forerunner of the RAT (Ram Air Turbine) which is fitted to modern aircraft to supply emergency hydraulics or electrics. It was possible with the 'H' type winch for the windmill to be removed and the winch operated from the normal aircraft hydraulic system.

The winch could be used for paying out or hauling in and its great advantage was the 7,000 ft (2,134 m) of 10 cwt (508 kg) cable which it could handle.

A pneumatically operated brake was controlled from the same control box which mounted the controls for the hydraulic motor. Motor control consisted of a valve which regulated the supply of fluid to the motor, thus controlling the speed and direction of the winch. This gave a very good low speed control which made target exchanging and recoveries very much easier than with other types of winch.

The winch was fitted with automatic laying-on gear, a footage counter and a cable speed indicator. Like the 'G' winch, it was intended for remote control use. The type 'H' was intended to be fitted to the Miles Monitor aircraft, but when it was decided not to continue with this aircraft, the winch was also discontinued.

Winch Type DX4 (Del Mar)

During the mid 'sixties, voices were clamouring for a long tow system to be introduced and a tow length of 60,000 ft (18,288 m) was being quoted as the ideal. There was nothing available in this country which could approach this figure, the 6,000 ft (1,920 m) of the 'G' winch and the 7,000 ft (2,134 m) of the type 'H' being the best at the time time.

Perhaps it would be as well at this stage to remind ourselves why this long tow length was deemed necessary. Much of the firing practice, both surface to air and air to air was changing from orthodox weapons to guided missiles which homed onto a target, by heat, light or radar. This meant that the target had to be fitted with an artificial heat, light or radar source, preferably bigger than that which the towing aircraft presented. These could be, and were provided, but the problem arose that if a missile was sent on its way and the artificial source on the target failed, the missile would then home onto the towing aircraft. It has been mentioned before that, when radar controlled guns were being used, great care had to be taken as it was not unknown for the guns to track up the towing cable to the aircraft. It was generally agreed that a tow length of 60,000 feet, almost seven miles (11.26 km), was the distance that provided absolute safety, hence the requirement for the long tow winch.

As noted earlier, attempts were made to uprate the faithful old 'G' type winch to carry the long tow,

65 Below left: A side view of the unlucky Miles M.33 Monitor, the only aircraft to enter British Service designed specifically in response to a 1942 requirement for target towing. The aircraft had to be capable of towing a target at 300 mph (483 km/h) at 20,000 ft (6,096 m) for 3-4 hours. It also had to be able to fly as slowly as 90 mph (145 km/h) while streaming a target. Originally intended to use the wing of the Beaufighter it was eventually the first aircraft from Miles which used a metal fuselage. The prototype, NF900, was lost in a fatal crash as the result of an engine fire, but in the event, the Royal Navy's needs were felt to be the most urgent and so it was never used by the RAF. By the time the machine was ready for service, the end of the war was in sight and only 20 were completed, 10 going to the navy, the remainder being broken up. This is NP406, the first production example.

66: *SN127 was one of 80 post-war Tempest TT Mk. 5 conversions. It had earlier served with both 56 and 33 Squadrons with the Occupation Forces in Germany in 1945, so this may be the location. It is finished in silver, black and yellow with yellow fuselage band.*

67: *A sketch showing the internal components of the Type 'G' winch, as used on Tempests and many other aircraft types post-1945.*

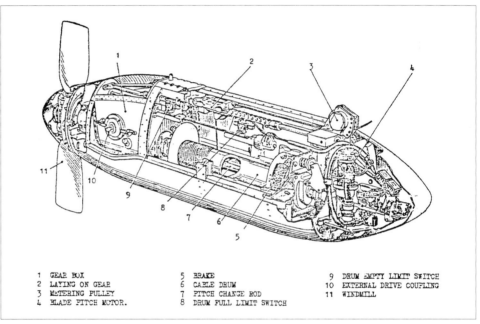

1	GEAR BOX	5	BRAKE	9	DRUM EMPTY LIMIT SWITCH
2	LAYING ON GEAR	6	CABLE DRUM	10	EXTERNAL DRIVE COUPLING
3	METERING PULLEY	7	PITCH CHANGE ROD	11	WINDMILL
4	BLADE PITCH MOTOR.	8	DRUM FULL LIMIT SWITCH		

68: *A Type 'G' winch in situ on the starboard wing root pylon of a Meteor TT Mk 20. Note that the aircraft has recently come in from the rain.*

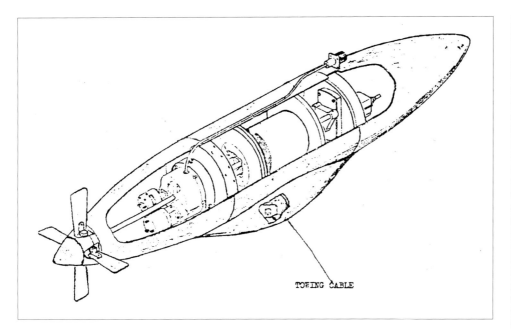

TOWING CABLE

69: *A sketch of the small American Del Mar DX4 long tow winch. Although only in service with the RAF and RN for a short time before it was superseded by the Rushton type, it gave the British Services their first experience of long tows (over 20,000 ft / 6,096 m of cable).*

70: *The Del Mar DX4 winch in position on the starboard wing pylon of Meteor TT Mk 20 WM234, sometime in 1966.*

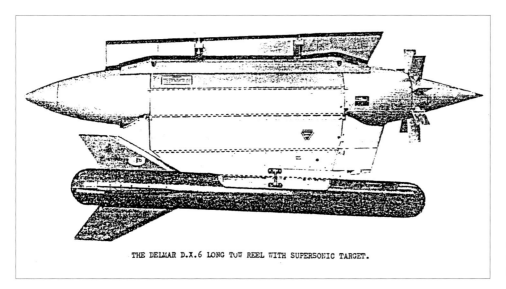

THE DELMAR D.X.6 LONG TOW REEL WITH SUPERSONIC TARGET.

71: *An indifferent diagram of the Del Mar DX6 winch, which was never used by British service aircraft, but saw extensive use with US Navy Phantoms and Skyrays.*

but this was never really a realistic proposition from the start. Consequently Flight Refuelling Ltd were charged with the task of designing and building a new long tow winch but as an interim measure it was decided to purchase a number of Del Mar DX4 winches from the USA.

The DX4 was an extremely well designed little winch which could handle 20,000 ft (6,096 m) of 0.051 inch (1.3 mm) mono-filament piano wire which had a breaking strain of 1,200 lbs (544 kg) which was considered adequate.

The DX4 was only half the size of the type 'G' and weighed only 350 lbs (159 kg) complete with cable. It was mounted on a pylon fitted to the upper surface of the mainplane of a Meteor TT Mk 20 aircraft in the same position as the Type 'G'. It was originally tried with the Del Mar target described in an earlier chapter. Later, the winch was fitted up in conjunction with the Rushton target and it was this combination which entered service with the RAF and Royal Navy, giving the British Services their first long tow capability.

The Del Mar DX4 Tow Reel was 6 ft (1.8 m) long and 1 ft (30 cm) in diameter. Perhaps it is as well mentioning that the Americans have always referred to their winches as 'tow reels'. As far as we are concerned, it is a winch. The winch was divided into three sections. The nose carried a four bladed windmill or air turbine to use more modern terminology, which had an adjustable pitch in both directions, the pitch being varied by a small electric motor contolled by a switch on the operators panel. The centre section contained the cable drum and laying on gear and the streamlined rear section housed the pitch change motor, the overspeed trip and the electrical distribution box.

The DX4 was unique in as much that the level wind gear did not move the feed pulley in the orthodox manner but moved the whole drum fore and aft to ensure the tidy laying on of the cable A fixed pulley allowed the cable to be pulled off the drum at right angles to the winch body.

A footage counter and an electrically operated disc brake was fitted. The maximum speed of the winch was 6,000 rpm and if this speed was exceeded, an overspeed trip would automatically apply the disc brake.

The brake was not intended to be used during winching and an interesting system was used. The winch was set to pay out at the maximum speed until the cable was within a few hundred feet of the required tow length. At this point, the pitch angle was decreased until it was set with a slight 'in' pitch. It was then adjusted so that the 'in' pitch was not enough to give power to haul in but enough to prevent the cable paying out. As soon as this point was reached and the windmill stopped, the brake was applied

After the Meteor TT Mk 20 had entered service with the DX4 winch and the Rushton Target, it was decided to equip a Sea Vixen aircraft with similar equipment. The winch was fitted to the underside of the starboard mainplane on the inboard pylon where a drop tank would have normally been carried, and the target launcher was fitted on a similar pylon outboard of the winch. The Sea Vixen / Del Mar combination proved quite successful during trials at A&AEE, but for reasons best known to the Ministry of Defence (Navy) it never entered service. The most likely reason was the imminent appearance of the Canberra TT Mk 10 with the new Rushton long tow winch.

The Del Mar DX4 had only a short life in the British services, but it represented an important stage in the development of the long tow system.

Winch Type DX6 (Del Mar)

The Del Mar DX6 was never used by the British Armed Forces, although at one stage it was being seriously considered for use with the Phantom aircraft as a supersonic target facility. At the time, the winch was being used by the American forces for this purpose in combination with the Hayes target on Phantom and Skyray aircraft.

To all intents and purposes, the DX6 was a scaled up version of the DX4 and was capable of handling 60,000 ft (18,288 m) of cable. The winch drum was driven by an eight-bladed air turbine which gave the winch the capability of winding cable in or out at speeds up to 5,000 ft (1,524 m) per minute with a turbine speed of 8,000 rpm.

The winch was normally used as a complete system pack with the target launcher being built on to the winch casing, using a Del Mar or Hayes target. The DX6 tow reel pack with the Del Mar target closely resembled the Rushton winch and target system in concurrent use with the Canberra TT Mk 18.

The Rushton Long Tow Winch

Quite the most sophisticated winch produced to date, the Rushton was used for many years with the Canberra TT Mk 18 aircraft and the Rushton Target. Designed and built by Flight Refuelling Ltd, the Rushton was originally designed to handle the magic 60,000 ft of tow cable, although in service, 35-45,000 ft (10,668 -13,716 m) was normally used.

These very long tow lengths presented tremendous problems and the Rushton has several novel features to minimise these. The heavy loads imposed on the cable due to the weight and drag of the cable with these long lengths tend to make the cable pull tightly on to the drum forcing strands to bed into each other which causes bruising of the cable. Inevitably, this causes breakage and was one of the main failures with the attempted uprating of the 'G' winch. To overcome this problem, the Rushton winch makes use of a capstan which hauls in the cable at a set rate and an inbuilt computer regulates the speed of the drum to take up the slack.

Another difficulty lies in the fact that the weight of the cable of this extreme length is probably more than the breaking strain of the cable. The Rushton winch therefore makes use of a stepped cable which starts on the target end with a thickness of 0.051 inches (1.5 mm), progressively increasing its diameter and strength until it reaches the drum, where a cable of 0.2 inches (5 mm) is fitted.

A three-bladed air turbine drives the winch through the medium of a pair of adjustable 'V' pulleys and belt which gives an infinitely variable drum speed. This enables the operator to select a speed on his panel which is sustained at all times during winching in or out.

The Rushton has many failsafe and safety devices which include an overspeed trip and an electrically operated explosive cable cutter. It will be appreciated that if a target weighing two or three hundred pounds is being winched in at two thousand feet per minute and is not stopped in time, it would strike the aircraft with considerable force inflicting serious damage. This contingency is catered for by a device which senses the cable speed, and if the cable is running too fast with the target too close to the aircraft, the cable cutter is automatically operated. The cable cutter can also be operated independently by the pilot or operator at any time should an emergency arise.

The Rushton winch incorporates a launcher for the Rushton Target making it a complete systems pack. Its many new features caused numerous initial problems during trials and early service life, but it is now a very proven winch which is possibly the most advanced in the world. For a considerable period only the Canberra TT Mk 18 was fitted with the system, carrying an independent unit under each wing, and this combination provided the British Armed Forces with an excellent target facility for many years.

It is interesting to note that although the Rushton system includes several highly sophisticated types of target, the winch also has the facility to operate banners and sleeves. (See photo on page 10).

*72 **Below:** Seen in May 1944 this is Defiant TT Mk I DR967, clearly showing the Type 'B' winch windmill and target stowage containers. Why it should be wearing a fighter-type Sky fuselage band is a mystery.*

*73 **Left:** An American Hayes launcher and target undergoing trials with a Meteor TT Mk 20 during the assessment phase prior to intended licence-building by Flight Refuelling Ltd. Various difficulties encountered ultimately led to the decision not to proceed, but to produce the home-designed Rushton winch and target system instead.*

ACCESSORIES
Hooks and cables

Over the years there have been many smaller parts, fixtures and fittings associated with the art of target towing, some specially designed and some AGS (Aircraft General Spares). It would be impossible to make a full list but it may be of interest to note some of the more important items such as release hooks, cable cutters, swivels and so on.

One of the most fundamental items would be release hooks as these were in use before the days of winches, when a target could be hooked up on to almost any kind of aircraft just by the fitting of one of these devices. Typical of these was the last of the purely mechanical types to be used in service, referred to as the Towing Hook Mk 2A. This device was still in use up to the end of World War II and was fitted to aircraft such as the Mohawk, Lysander, Harvard and undoubtedly many other types. It was normally bolted to a strong point on the underside of the aircraft and was was operated by the pilot through a Bowden control cable.

A number of electro-mechanical releases were available to replace the Mk 2A, notably the ML 3,000 lb release slip. This type has been around for several years and is still available today. It was fitted to the Meteor TT Mk 20 when using the Brooklands Dart Target and also to the Sea Hawk, Scimitar and Canberra aircraft using the air snatch technique. It is a very simple but reliable release and is usually wired so that it can be used by the pilot or operator.

Another well used type was the AST Twin release. This was not unlike the ML in operation but was basically two release units side by side. This was also used by a variety of aircraft, but I saw it fitted to a Corsair aircraft. The slip was mounted on the centre section and a folded sleeve target was carried in a cannister under each wing, each cannister also carrying about 1,000 ft (305 m) of nylon tow line. The tow lines were brought across to the centre of the fuselage and engaged in the jaws of the twin towing slip. The first sleeve was deployed and after it was damaged during air firing, it was released and the second one deployed. This system was used for air-to-air firing.

A very necessary fitment to all winch-equipped aircraft is the emergency cable cutter. Like the release slip, the earlier ones were purely mechanical in operation, usually by the use of a Bowden cable with a release handle in both pilot and operator's cockpit.

Three typical mechanical cutters in service were the ML type which was fitted to the Meteor TT Mk 20 and the Mosquito TT Mk 35; the Blackburn and General Aircraft type as fitted to the Beaufighter TT Mk 10 and the Sturgeon TT Mk 2, and the Hawker which was fitted to the Tempest TT Mk 5.

All of the mechanical cable cutters were similar in design and operation, namely a heavily spring-loaded plunger with a hardened steel cutting blade which operated against a steel anvil. The plunger was held back against the spring by the cable operated catch and the towing cable was passed between the plunger and the anvil. When the catch was operated, the spring forced the cutting edge on to the anvil thus cutting the cable. These mechanical types were

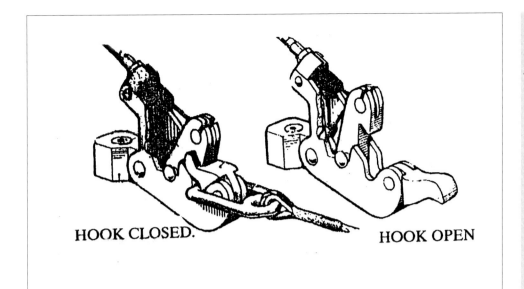

HOOK CLOSED.   HOOK OPEN

75: *This is an accessory; namely the Towing Hook Mk 2A in both open and closed positions.*

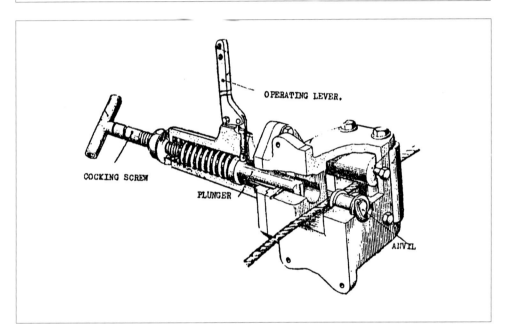

OPERATING LEVER.

COCKING SCREW

PLUNGER

ANVIL

76: *Here is the Blackburn and General Aircraft Emergency Cable Cutter; a mechanism of last resort.*

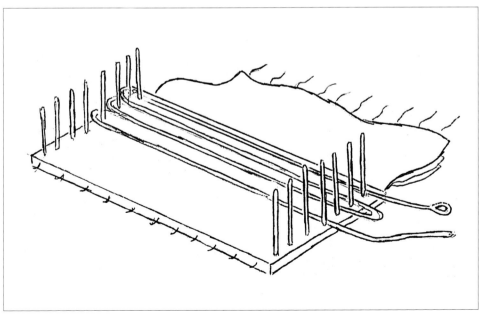

77: *This sketch shows the rope dispenser contrived to allow a Canberra to safely pick up a 6,000 ft tow rope by the snatch technique.*

loaded by compressing the spring with a cocking screw which was removed after the release catch engaged with the plunger. The screw was usually left in when the aircraft was on the ground and removed at the same time as the control and undercarriage locks prior to flight.

Later types of cable cutter dispensed with the spring and used an explosive cartridge instead. The cartridge, which is fired electrically, drives a piston with a cutting edge attached on to the anvil, cutting the cable in the normal way.

These explosive cable cutters are usually used when piano wire is being employed as a towing cable although trials have shown that they will also cut stranded 15 cwt (762 kg) cable equally as well. Apart from the obvious advantage of being able to dispense with the long run of Bowden cable, the explosive cable cutter can be considerably reduced in size and weight. Modern winches such as the Rushton incorporate the emergency cable cutter in the winch pack.

Until the introduction of monofilament piano wire, the tow cable which had been in use for many years was the type known as 'Kelindo'. This was available in two sizes, 15 or 20 cwt (762 or 1,016 kg), the former being most common in use. Before the introduction of Kelindo, ordinary flexible hawser laid cable was used but this often suffered from breakage due to 'basketing'. This was caused by a tendency for the cable to rotate under load and unwind. Kelindo was introduced as a non-rotating cable which was laid up in a different way to normal. It consisted of an inner cable which was wound in a clockwise direction and an outer layer which was wound anticlockwise. Any tendency for the cable to twist would make the inner and outer cables to work against each other thereby resisting any tendency to rotate.

When the SAAB Near Miss Recording System was introduced, another type of cable had to be used. The SAAB cable was similar to the Kelindo, but had an insulated heart strand running through the centre.

The complex nature of the construction of the SAAB and Kelindo cables made it very difficult to splice. Although there was seldom a requirement to splice the two ends together, it was important that an eye could be formed in the end of the cable for attachment to targets. The Air Publication on Target Towing, (AP 1492) gave full instructions for splicing but it was such a long and tedious process that it was seldom carried out.

To make matters easier, the Talurit Swage was introduced. This consisted of an aluminium ferrule which was threaded on to the cable which was then fed round and back thjrough the ferrule to form an eye. The ferrule was then crushed between the jaws of a hydraulic press. The Talurit Swage has been in common use throughout industry for many years and has proved to be the most effective way of forming an eye in the end of a cable. Once formed, it will take a load up to the breaking strain of the cable.

A target, particularly a sleeve, would often have a tendency to rotate along its longtidudinal axis when being towed through the air. This rotation is not necessarily a bad thing; in fact it does to a certain extent improve the stability of the target. However, it is easy to imagine that this rotation would soon twist the cable up into a hopeless mess. This was easily remedied by the fitting of a swivel between the target and the cable which allows free rotation of the target independant of the cable. The swivel in current use has a ball bearing thrust race to allow easy rotation and has a safe working load of 3,000 lbs (1,360 kg).

When talking of cables, it is worth mentioning leader cables. This is a comparatively new idea. It was found that when using the small diameter single strand piano wire, problems arose when recovering the target back on to the aircraft. When the target was close hauled it was badly affected by the slipstream of the aircraft and would often yaw violently from side to side causing the piano wire to snap. To overcome this problem, a 25 ft (7.6 m) length of 10 cwt (508 kg) flexible steel cable was fitted between the target and the end of the piano wire. This stronger and more flexible cable was able to hold the target during the critical recovery period. The problem arose in tying to join the two cables together. Eventually, it was found that the piano wire was slightly larger than the central heart strand of the hawser laid cable. By unlaying the cable and removing a 6 ft (1.8 m) length of the. heart strand, it was possible to replace the heart strand with the piano wire. A good whipping at the end of the flexible cable made a very strong joint which was up to the total strength of the wire.

During the Canberra snatch programme, there were occasions when a 6,000 ft (1,830 m) tow length was required. A rope dispenser had to be used to contain 2,000 ft (610 m) of tow line as 4,000 ft was the maximum length that could be laid out due to the size of the snatch area.

The rope dispenser consisted of a wooden platform measuring 4 ft (1.2 m) by 2 ft (0.6 m), with a row of holes either end in which $1/4$ in (6.5 mm) steel rods were inserted. A pack of fabric sheets was attached to one side with strings attached to the outer edge which could be tied to hooks on the other side of the board. When packing, the rope was laid in as shown in the sketch. With the first layer laid on, a fabric sheet was pulled over and tied off on the hooks. The following layers were laid on in the same fashion with a fabric sheet between each layer. When the full 2,000 ft (610 m) of rope was laid on and the fabric sheets tied off, the steel rods were removed and the dispenser was ready for use.

78: A rope eye formed by the Talurit swage Method — read on!

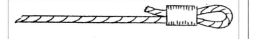

A smaller dispenser, using a similar method was used with the aerial Banner release which contained 800 ft (244 m) of rope. These dispensers allowed rope to be pulled out at around 2,000 feet per second without tangling.

With the Mk 3 Dart installation on the Meteor TT 20 a small fabric dispenser was filled between the two upper fins as shown in the sketch opposite. This was to carry the 20 ft (6 m) of nylon rope on which the dart was initially launched.

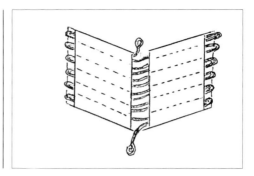

79: A sketch of the fabric dispenser fitted to the two upper fins of the Dart Mk 3 Target installation on the Meteor TT Mk 20.

80: Another view of Wallace Mk II K6015 in 1936 while it was in use with the Anti-Aircraft Co-operation Flight at Biggin Hill. From this angle it can be clearly seen that it is fitted with a Type 'B' winch, the cable from which emerges under the fuselage. It is prevented from fouling the tail surfaces by various guards.

81: Battle TT L5598 underwent tests at Boscombe Down in May 1940 as a target tug; a role for which it was far better suited than its operational compatriots across the Channel in France at the same time. The Type 'B' winch is in the stowed position and there is some wrinkling on the fin. L5598 never went to an active unit before being scrapped.

82: This mysterious picture shows three Fairey IIIFs, all with dark camouflage and Type 1 British roundels with white or yellow stripes on the wings, tails and fuselages. The style of markings suggests the period could be about the time of the Munich Crisis in 1938. The nearest machine appears to have a sleeve target draped around the port wing strut, and a Type 'B'(?) winch is visible projecting to port from the rear cockpit. The strong light suggests a warm location, but where? Malta? The Middle East? Or just a fine day in the UK?

83 Above: *How to spoil the lines of a beautiful aircraft. This is Mosquito TT Mk 39 PF 606 in September 1948. Originally a Percival-built B Mk XVI, this was one of a batch converted for target towing duties with the Royal Navy and is seen here in immaculate condition just after rollout at Feltham. Fitted with an ugly birdcage nose and festooned with numerous guards and bumps and many stencil markings, PF606 was actually mostly used for radio trials at Lossiemouth until it was scrapped late in 1952.*

THE AIRCRAFT
Unsung warriors

It seems probable that well over a half of the aircraft types used by the British Armed Forces since 1930 have, at some time in their life, been used for target towing in one form or another.

In most cases, aircraft have been obsolete before being relegated to this role; certainly, this has been the case with the majority of the specialised target towing aircraft. Generally, the towing aircraft can be divided into three categories as follows:

1:   In this group are the basically standard operational aircraft, usually fighters such as Scimitar, Hunter and Javelin, which were adapted with minor modification for the towing of banners or sleeves with a fixed tow length. These modifications normally consisted of little more than the fitting of a towing hook for the attachment and release of the target.

The aircraft in this group were rather restricted to this type of work as the tow length was comparatively short, up to 1,000 ft (305 m), because of the fact that the aircraft had to drag the target off the ground, which limits the take off run. This short take off limited the use to air-to-air firing using conventional cannon or machine guns. This system was normally used by the Squadron or Gunnery School concerned. Today, however, the inbuilt weapons carrying capability of most fighter aircraft, such as the Hawk, and the small size of the target equipment allows them to readily carry out target towing duties.

2:   In this category are bomber aircraft adapted for the training of air gunners by removing the rear turret and fitting a hand winch.

3:   These were the aristocrats of the target towing world and consist of aircraft which have either been extensively modified or specially designed for the task being fitted with winches, target launch facilities and other special fitments necessary for the aircraft to carry out its target towing role.

Very few types have been specifically designed for target towing, the most notable exception being the Miles Monitor, which, by having made history by this unique distinction, was doomed never to enter service. The Miles Martinet, probably the most famous of all target towers, was only a development of the Miles Master trainer. Some aircraft which failed to make the grade in their original design were converted as target towers, these include the Hawker Henley dive-bomber, the Fairey Battle bomber and the Short Sturgeon torpedo-bomber.

The majority of tugs were aircraft which had become obsolete in their original role and were given a final fling as target towers. These include the Fairey Firefly, Bristol Beaufighter, De Havilland Mosquito, Gloster Meteor and latterly, the BAC Canberra. Most of these ended up with the Royal Navy's FRADU (Fleet Requirements and Air Direction Unit), from which the last was retired in 1992.

For many years, target towing aircraft were identified by the painting of the underside of the machine with black and yellow diagonal stripes with the upper surfaces being more or less standard camouflage. In later years it was the practice to paint the upper surfaces silver with bands of 'Dayglo' red, although

Hawks in the target-tug role continue to wear standard low-visibility finish.

Another modification which featured on tug aircraft was the fitting of safety wires. These were fitted from the top of the fin to the outboard tips of the tailplane, then down to the underside of the fuselage tail. They were meant to prevent the towing cable being caught up in the tail. The Meteor TT Mk 20 started life with these cables, but they were soon removed as being unnecessary. The Canberra never had them. The cables were obviously an embarrassment to modern high performance aircraft but one wonders why they were considered no longer necessary. Perhaps modern tow cables are better trained, but I remember a Meteor causing us an anxious half an hour before it finally came in and landed safely with a length of tow cable wrapped around the elevator.

Before the war, there were no target towing squadrons as such and most of the target flying was carried out by special service flights attached to the squadron or by the squadrons themselves. During the war, however, there were about half a dozen Tar-

get Towing Squadrons, the author being a member of 286. After the war, although the RAF equipped 7 and 100 Squadrons with Canberra TT Mk 18s, most of the target flying was carried out by civilian contractors such as Airwork and Flight Refuelling Ltd (today Cobham Plc).

In more recent years the preferred option by the Ministry of Defence has been for unmanned drones; either conversions of obsolete types such as the Firefly and Meteor or specially developed UAVs (Unmanned Aerial Targets). Predominant among these are the Australian-developed 'Jindivik'. This has been the standard weapons target for the RAF fighter squadrons since the early 1960s, but is to be replaced by the Italian-designed Mirach.

Target towing by manned aircraft in Britain, however, continues, as Hawks of 100 Squadron now act as replacements for the venerable Canberra.

It would be a massive task to list all of the aircraft which have been involved with target towing, but in the following pages we shall see some of the types that have been associated with this menial task.

84: One of few known pictures of Lysanders in service with an actual target towing unit. This is Mk II P1688 while in use with 3 Armament Practice Camp at Leuchars, evidently sometime after June 1942 when the later type national markings seen here came into effect. The aircraft was scrapped on 25 August 1943.

85: By way of contrast, this is Hawk T Mk 1A XX177, one of the earliest built, seen in 1995 in service at Valley with 100 Squadron. Modern aircraft have more than adequate power to pull long tows and are easily fitted with bolt-on attachments, as with the target pod under the fuselage here. Current targets are not only used for gunnery practice, but also for radar calibration and similar tasks. Note the low-visibility finish and the unit's skull and crossbones emblem.

86: *One of the more exotic types to be used for target and gunnery purposes was this Boulton Paul Overstrand, K4563. This was one of several which had once belonged to 101 Squadron before eventually ending up with 10 Bombing and Gunnery School where it was pictured, here, in 1940.*

87: *Henley Mk II L3276 was one of three used at Boscombe Down. L3276 was used, for some odd reason, in trials as a bomber in January 1942. It retains the target towing finish, although the winch has been removed. It eventually went to 1628 Flight as a tug.*

88 Below: *Henley L3247 was used for nine months from September 1939 in its intended role as a target tug at the A&AEE. It appears to have kept its camouflage on the horizontal tailplane and port wing tip. Note the black stripe overlapping the starboard wing roundel and the diamond gas marking on the rear fuselage spine.*

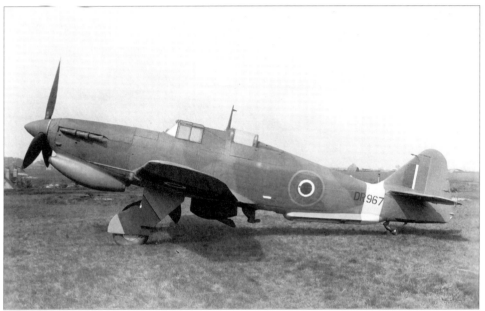

89 Above: *Defiant TT Mk III prototype DR863 undergoing tests at Boscombe Down in July 1942. The stowage containers for the sleeve or banner targets are fitted on the lower fuselage.*

90 Left: *This port side view of Defiant Mk I DR967 shows clear details of the stowage containers, the pulleys for the cable feed and the various guards around the tail surfaces. The rearward facing winch operator's position is also visible. A huge fairing over an enlarged oil cooler has also been fitted under the nose, presumably in connection with engine overheating generated by the extra drag of the targets.*

91: *Martinet prototype LR244 undergoing performance trials at the A&AEE in 1942. The towing cable arm can be seen projecting below the fuselage, but apart from that it shows no other evidence of its intended role. 1,724 Martinets were produced, including 65 radio-controlled drone Queen Martinets, yet only a handful of photos exist...*

92: *Pictures of target tugs in unit service are rare. Here, however, is Martinet TT Mk I HP200 while in service with 595 Squadron at Pembrey in 1947, still in wartime markings and camouflage.*

93: *After valiant service in the early part of the war, more capable aircraft replaced Blackburn Skuas. Surplus examples found use as target tugs. Wearing immaculate black and yellow stripes overall, L3007 deserved better — it shot down three Italian aircraft between 1939 and April 1941, when it was converted to a target tug and allocated to 757 Squadron at Worthy Down, probably about the time this picture was taken. It survived service with three other Fleet Air Arm squadrons until May 1944.*

94: *An example of the Airspeed Oxford Mk I, intended as a dedicated gunnery trainer. Consequently, Oxfords often operated in conjunction with target towing aircraft. This one, N6250, when seen here was in use with 2 FTS in 1940, but eventually ended its service with 1624 (Anti-Aircraft Co-operation) Flight in October 1943.*

95 Above: *The portly Brewster Bermuda was not a success. FF557 was tested by the A&AEE as a target tug in November 1943 with an 8 ft banner target on a 2,000 ft tow, streamed from a Type 'B' winch, which tended to overheat. A smaller target was recommended.*

96 Left: *Harvards were not often used as target tugs, but this one, based at Luqa, Malta, did so postwar.*

97: *As the need for faster moving targets grew, so obsolescent or worn out fighters found a new lease of life with target-towing units. Just a very few were Mustang Mk Is, similar to AG351 here. Four definitely known to have served in this way were AG386 (to 286 Squadron after Maclaren swivelling undercarriage trials); AG561 (1471 AACU Flight); AG618 (1483 Target Towing & Gunnery Flight) and AP168 (285 Squadron). As a general rule, however, Hurricanes were found to be less useful as fighters and allocated to this role.*

98: *Post-war, more potent fighters found use as tugs, including a number of Tempest Mk Vs converted to TT Mk 5s. Among these was SN261, seen here serving with the Central Gunnery School in 1949. The cable arm for the tow cable can just be seen below the central fuselage, while the aircraft displays typical target tug finish. The full code is not known, but was probably FJU-?.*

99: *Miles Monitor NP407, the second production aircraft, on its maiden flight on 5 April 1944 in the hands of Flt Lt Tommy Rose. The excellent view for the winch operator is self evident.*

100 Below: *Mosquito TT Mk 39 prototype PF489 in August 1947. It appears to have some kind of fabric or plastic covering over the glazed nose. Before conversion from a B Mk XVI, it had been used by 105 and 608 Squadrons. As a tug it went to the Royal Navy after handling trials at the A&AEE which criticised the view and its handling.*

101: *The ungainly Short Sturgeon TT Mk 2, originally intended as a Far East torpedo-bomber. As modified it could tow 16 ft and 32 ft wing targets as well as various sleeve and banner types. Accepted for RN service in early 1950, it was easily capable of serving aboard aircraft carriers of the time, but by the mid-'fifties most were surplus to requirements and placed in store prior to scrapping. Note the break line on the nose which allowed it to be folded to save space on the aircraft carriers.*

102: *VR371 was the second prototype Sturgeon Mk 2. Most of the target towing equipment was concealed within the capacious fuselage, although a robust cable guard can be seen in front of the tailwheel.*

103 Below: *A Firefly TT Mk 4 with a Type 'G' Mk 3 winch nestling between its legs. WB406 carried out trials at the A&AEE in late 1950-51 with a 32 ft winged target, favoured by the Admiralty.*

104: *Many aircraft types were tested for their suitability to tow targets in the 1950s, including a Javelin FAW Mk 6, XA821, similar to the one shown here, which pulled a 6 ft x 30 ft banner at the A&AEE in 1959.*

105 Below: *One of the earliest drone targets was the Airspeed Queen Wasp, the landplane prototype, K8887 being seen here in early June 1937. Despite being intended as a radio-controlled target, it was only destroyed by a wartime air raid on the Airspeed works at Portsmouth. Six or seven were built.*

106: *WM810 was one of the later family of drone targets developed from the Fairey Firefly Mk 7. Known as the U Mk 8, this was the first of its kind, seen in December 1953. It remained capable of being flown manually. The wingtip pods contained small cameras to track the movement of the aircraft and incoming missiles. At the time they were classified as 'Top Secret'. All the drones were finished in this striking red and yellow colour scheme.*

108: *This is the third Firefly U Mk 8, WM856, without human pilot.*

109 Below: *Jets became drones also. This is a Meteor U Mk 16, flying under ground control from Aberporth in the early 1960s. WH344 was originally built as an F Mk 8 and served with 504 Squadron before conversion to a drone in April 1960.*

111 Above: *The recovery team at the scene on the Lulworth Ranges after the first flight of the A&AEE Trident target in 1961. From left: 'Wilbur' Wright; Colin Prouten; the author; Jock Paton. The Trident proved so successful that Don Evans and the other two members of the team were awarded the British Empire Medal for their efforts. Note the condition f the target.*

110 Left: *Backbone of the UK's pilotless target force for some 40 years was the Australian-designed Jindivik (an Aboriginal word meaning 'the hunted'). This one is in flight off the Welsh coast on 28 November 1978. © Crown Copyright.*

TALES AND TAILS
Unsung heroes

The previous chapters have been devoted to the hardware used over the years in target towing, but even more important of course is the men who have used the equipment. In the author's experience, most people that I have known have been most upset when they learned they were being posted to a Target Towing Squadron. Yet, although they would rarely admit it, after a few weeks have found themselves liking the experience. Looking back in later years they generally agree that it was one of the most enjoyable parts of Service life. That was certainly my reaction and also that of many I have met who have spent some time engaged in what the uninitiated regard as a most dreary pastime.

Every tug pilot or operator will have a few tales to tell, mostly amusing, some grim, mostly true, but a detailed account of the men who fly through the air like Bo Peep's sheep, bringing their tails behind them, would fill a volume.

Some of these stories are purely legend, the most famous being the knot in the cable. I have met at least three people who knew someone who had met a man who was told about a member of their Squadron who had deliberately tied a knot in the towing cable by pulling the aircraft up into a loop and rolled round the cable on the way out. Of course, a knot in

the cable will always cause the cable to snap, and in this individual's case, when the cable snapped the knot was on the wrong side of the break and was lost with the cable, leaving him with no evidence of his remarkable feat.

Wartime gunners, especially in anti-aircraft gunnery training schools, were notoriously inaccurate and it was not unusual to have flak bursting around the tug aircraft. This situation gave rise to the legendary shouts from the tug pilots, such as: *"Hey, I'm the one in front"*, or *"I'm pulling this bloody thing — not pushing it"*.

I remember on one occasion, while sitting in the back seat on a Martinet putting up targets for a gunnery school near Chesil Beach. It was customary on these runs for the pilot to call up the gunners to check if they were ready, and if they were, inform them he would do a dummy run during which the target would be streamed and paid out, followed by a reciprocal run for firing. On this particular trip, the usual formalities were observed and the people below said they understood. The pilot told me to launch and stream the target which I duly did, and with the sleeve trailing behind on a 50 ft (15 m) long halyard, I was about to set the winch paying out. No sooner had I put my hand on the brake lever than all hell was let loose. Those clever soldiers were firing at the target and flak was bursting all around. The message that my pilot shouted to the gunners could not be repeated in these pages.

On another occasion, I was again in the back seat of a Martinet, this time putting up targets for a battery of naval 3.7 inch guns at St Catherines Point on the southernmost tip of the Isle of Wight. We had carried out several runs against some pretty good shooting, having had to exchange a couple of sleeves. We were on the final run from north to south and were just clearing the range, the guns had ceased firing as we were out of range and we had just started a slow turn back to base at Weston Zoyland in Somerset. Suddenly, some hefty looking tracer started shooting up past the port mainplane. The usual profanity from the pilot to the matelots down below produced the reply that it wasn't them. Not only were we out of range, we were out of sight. *"Get rid of the target for Christ's sake!"* the pilot shouted at me. *"Look down there!"* Straining to look over the side, I was most upset to see a German 'E' boat pumping everything he had at us. Thank God he wasn't such a good shot as the matelots we had been playing with earlier! We weren't fighting men, so we cut the cable and hurried home.

Some of us were not so lucky. I lost three friends. One was lost when one of our Martinets came down in Lyme Bay, possibly due to aircraft failure but more likely to bad shooting. Friendly fire they call it these days. Another was in a Defiant which made a wheels-up landing at Weston Zoyland, but unfortunately cartwheeled and caught fire, and the third had half his arm ripped off in the back seat of a Henley due to a cable snarl up. He was alive when the aircraft landed but died in hospital later through shock, loss of blood or something pretty final. Ted Yates's death was particularly sad as we had attended his wedding in Bridgwater only a few weeks earlier.

It was a very funny thing, but if you were flying in a Defiant which was built mainly from metal, you seemed more secure. The Martinet was mainly wood and you felt very exposed. Of course, it was all in the mind. If a lump of flak hit you in the wrong place, you were down no matter what you were flying in.

Despite the darker moments, I look back to my days as a winch operator as one of the most enjoyable in my life. I was in fact an engine fitter, but like several in our Squadron, was enticed by the princely sum of ninepence a day to hook on a four thousand foot tail and go flying. It was not just the money. We were able to lounge around the crew room when we weren't flying, wearing a big pair of wellington boots and a Mae West. The WAAFs thought we were great. So did we. We were *aircrew*.

Seventy-six sorties, a four and a half hour marathon in a Vengeance and a wheels-up landing in a Defiant later, my posting finally came through. I thought it was goodbye to target towing, but, years later, I found myself in charge of a team involved with Towed Target Development at Boscombe Down. Not so much flying for me, different machines, different targets, but the laughs were still there. During the course of my duties I had occasion to visit several Service units and found the same sort of spirit still existed in the 'mob' — everyone moaning about being stuck on target towing, but enjoying every minute of it.

One of the most exciting and amusing periods was during the spectacular air snatch trials. Many of these were carried out at the Army Tank Ranges at Lulworth. Here we had our own tank to set up and recover the targets. On one occasion I had fired a green Verey cartridge to give the pilot the all-clear for the drop, when the spent cartridge, still very hot, fell to the ground and set light to the ranges. I had to tell the pilot to drop the target as best as he could as we had a fire to deal with. The fire proved to be more that we could cope with, and two and a half hours later, half of the Dorset Fire Brigade and a third of the British Army were involved. The Brigadier gave me a right old dressing down and it was more than I dared admit that I was also a part-time fireman.

On one occasion at Lulworth, a soldier who was an interested spectator, and close enough to see the aircraft and target, but not close enough to see the towing cable, congratulated us on having such a wonderful missile which took off and followed the aircraft, then continued to follow it until we told it to let go.

From time to time, targets were lost over land and it was my job to set off and make friendly noises to some farmer or other to recover the target. Most of these were happy enough, especially when we forgot to take a few hundred feet of nylon tow cable away with us. Very useful stuff to have around the farm.

When these targets were lost over land they were often difficult to find, and one of our pilots (who later became a high-ranking officer in the RAF) introduced a great recovery service. He would send us off in our 4-ton truck, which was equipped with all colours of radio, in the general direction of the truant target. He would then take off in a Meteor after giving us sufficient time to reach the area, then fly round us giving the appropriate directions: *"Turn left at this crossroads. Through the second field gate. Across the ploughed field. Left a bit. There you are. You've got it".*

On one of these occasions we were operating from Everleigh Dropping Zone, just to the north of Tidworth, with a particularly troublesome rapid target exchanger. The target broke off and fell away, landing on the other side of a wood which obscured our view. We immediately set out with the truck on a recovery mission under the guidance of the aircraft. *"Back through Everleigh village and turn left"* he sang out. *"It's lying by the side of the road on a bend about half a mile ahead of you"*, was his next message, quickly followed by another: *"Hurry up. There's a van stopped and a bloke is stuffing our target in the back. Now he is moving away. I'll try and follow him"*. With my foot hard down on the floorboards we set off in chase and still guided by the pilot, reached Netheravon. *"Can you see the big white house with the green gate?"*, said the pilot. I replied in the affirmative. *"Well, that's where the bloke with the van went so I will leave the rest to you"*. The 'bloke with the van' was most surprised when we hammered on his door. *"Please can we have our target back... ?"*

These are only some of the tales I could tell. There are many others, such as the tractor driver ploughing a field near Marlborough who had had a 150 lb (68 kg) dart target whistle into the ground no more than ten feet (3 m) from the front of his tractor. When we arrived he calmly assured us: *"Don't worry mate. It was miles away"*. Then there was the Army colonel riding his horse across the ranges who refused to move away when the aircraft was approaching for a target drop, and whose horse finally threw him when the target landed a few feet away.

Tales too of the happy hours spent in Lyme Bay on an Air Sea Rescue launch, hove to waiting for a drop with the mackerel lines over the side.

Other people also have their tales to tell. At Hurn they talk of the two pilots who were racing each other to drop their targets before landing. Both dropped their sleeves at the same time, with one dropping his on top of the other's winch, where it tangled around the windmill and snapped it clean in two.

The Exeter crews like to tell of the time when one of their Mosquitos had operated its cable cutter after losing a target at Manorbier Ranges. Not realising that the cutter had not operated they came in for a normal landing at Exeter Airport. A low approach over that fair city with 6,000 ft (1,830 m) of cable hanging out from the back is not to be recommended. Half the lights of Exeter were put out, including those at the local hospital, countless chimneys were amputated and questions were asked in the House.

At Exeter, aircraft returning with targets on a short haul were subsequently dropped on the airfield where they were picked up by the fire tender. The Fire Chief, a bad tempered five-footer, used to complain bitterly if the target was dropped too far away from the crash bay. The aircrew, only too willing to oblige, tried to drop their sleeves straight through the doors of the Fire Station and were successful on more than one occasion.

It may sound as if target towing was a series of incidents, good and bad. Far from it. Most of the time is spent cruising up and down a range with little to break the monotony. One could go on for ever reminiscing, but this was not the intention when producing this survey. I can only apologise and hope that the human side has made it easier to read. All I can be sure of is, as long as we have weapons, we shall need targets to practice with, and as long as we have targets there will be men to operate them, and the men who tow tails will have tales to tell.

112 Right: *A rear view of a Rushton winch on a Canberra TT Mk 18 carrying two sleeve targets.*

113 Below: *A well worn Vultee Vengeance Mk IV, FD243, tested at Boscombe Down in April 1944. It is believed to have been later converted to target towing status and used by the author's own unit, 286 Squadron, at Weston Zoyland in late 1944.*

THEY ALSO SERVED

A selection of known UK-based Royal Air Force Target Towing Units

The Target Towing Flights were mostly formed in October 1941 from the previous Group Target Towing Flights

1479 (Anti-Aircraft Co-operation) Flight, Peterhead 1942
1480 (Anti-Aircraft Co-operation) Flight, Ballyhalbert 1941
1481 (Target Towing) Flight, Binbrook 1941
1482 (Target Towing & Gunnery) Flight, West Raynham 1941
1483 (Target Towing & Gunnery) Flight, Newmarket 1941
1484 (Target Towing) Flight, Driffield 1941
1485 (Target Towing) Flight, Coningsby 1941
1486 (Target Towing) Flight, Valley 1941
1487 (Target Towing) Flight, Warmwell 1941
1488 (Target Towing) Flight, Shoreham 1941
1489 (Target Towing) Flight, Coltishall 1941
1490 (Target Towing) Flight, Acklington 1941
1491 (Target Towing) Flight, Inverness 1941
1492 (Target Towing) Flight, Weston Zoyland 1941
1493 (Target Towing) Flight, Ballyhalbert 1941
1494 (Target Towing) Flight, Long Kesh 1941
1495 (Target Towing) Flight, Sawbridgeworth 1942
1496 (Target Towing) Flight, Hawarden 1942
1497 (Target Towing) Flight, Macmerry 1942
1498 (Target Towing) Flight, Hurn 1942
1499 (Bomber) Gunnery Flight, Wyton 1943
1600 (Anti-Aircraft Co-operation) Flight, Weston Zoyland 1942
1601 (Anti-Aircraft Co-operation) Flight, Weston Zoyland 1942
1602 (Anti-Aircraft Co-operation) Flight, Cleave 1942
1603 (Anti-Aircraft Co-operation) Flight, Cleave 1942
1604 (Anti-Aircraft Co-operation) Flight, Cleave 1942
1605 (Anti-Aircraft Co-operation) Flight, Towyn 1942
1606 (Anti-Aircraft Co-operation) Flight, Bodorgan 1942
1607 (Anti-Aircraft Co-operation) Flight, Carew Cheriton 1942
1608 (Anti-Aircraft Co-operation) Flight, Aberporth 1942
1609 (Anti-Aircraft Co-operation) Flight, Aberporth 1942
1611 (Anti-Aircraft Co-operation) Flight, Langham 1942
1612 (Anti-Aircraft Co-operation) Flight, Langham 1942
1613 (Anti-Aircraft Co-operation) Flight, West Hartlepool 1942
1614 (Anti-Aircraft Co-operation) Flight, Cark 1942
1616 (Anti-Aircraft Co-operation) Flight, Martlesham Heath 1942
1617 (Anti-Aircraft Co-operation) Flight, Newtownards 1942
1618 (Anti-Aircraft Co-operation) Flight, Cleave 1942
1620 (Anti-Aircraft Co-operation) Flight, Bodorgan 1942
1621 (Anti-Aircraft Co-operation) Flight, Aberporth 1942
1622 (Anti-Aircraft Co-operation) Flight, Gosport 1943

1623 (Anti-Aircraft Co-operation) Flight, Roborough 1943
1624 (Anti-Aircraft Co-operation) Flight, Detling 1943
1625 (Anti-Aircraft Co-operation) Flight, Weston Zoyland 1943
1626 (Anti-Aircraft Co-operation) Flight, Langham 1943
1627 (Anti-Aircraft Co-operation) Flight, Ipswich 1943
1628 (Anti-Aircraft Co-operation) Flight, Morfa Towyn 1943
1629 (Anti-Aircraft Co-operation) Flight, Hutton Cranswick 1943
1630 (Anti-Aircraft Co-operation) Flight, Acklington 1943
1631 (Anti-Aircraft Co-operation) Flight, Shoreham 1943
1632 (Anti-Aircraft Co-operation) Flight, Montrose 1943
1634 (Anti-Aircraft Co-operation) Flight, Hutton Cranswick 1943
1677 (Target Towing) Flight, Netheravon 1944

Several of the Anti-Aircraft Co-operation Flights were formed from the eight short-lived RAF Regiment Anti-Aircraft Practice Camp Target Towing Flights

1 RAF Regiment School Target Towing Flight, Hutton Cranswick 1942
3 RAF Regiment School Target Towing Flight, Ronaldsway 1942
Target Development Flight, Oakley, 1944
Target Facilities Flight, Binbrook 1966
Target Facilities Flight, Leuchars 1966
Target Facilities Flight, Wattisham 1966
285 Squadron, Wrexham 1941
286 Squadron, Filton 1941
287 Squadron, Croydon 1941
288 Squadron, Digby 1941
289 Squadron, Kirknewton 1941
290 Squadron, Newtownards 1941
291 Squadron, Hutton Cranswick 1943
1 Group Target Towing Flight, Goxhill 1941
2 Group Target Towing Flight, West Raynham 1940
3 Group Target Towing Flight, Marham 1940
4 Group Target Towing Flight, Driffield 1940
5 Group Target Towing Flight, Driffield 1940
6 Group Target Towing Flight, Abingdon 1938
9 Group Target Towing Flight, Valley 1941
10 Group Target Towing Flight, Warmwell 1941
11 Group Target Towing Flight, Shoreham 1941
12 Group Target Towing Flight, Coltishall 1941
13 Group Target Towing Flight, Acklington 1941
14 Group Target Towing Flight, Inverness 1941

Note: *Places and dates shown are those when the unit first formed. This list is far from complete — a fully comprehensive history of all the target towing units remains to be written.*

114 Below: *Towing an appropriately marked sleeve target, this was the final flight of Lincoln B Mk 2 RF533 on 21 April 1967. Fitted with an extended nose and extra day-glo panels to its standard post-war bomber camouflage, the aircraft had served for 13 years with the RAE at Farnborough on numerous trials, but not including target towing!*